Physico-Mathematical Theory of High Irreversible Strains in Metals

Physico-Mathematical Theory of High Irreversible Strains in Metals

V.M. GRESHNOV

CISP

CRC Press is an imprint of the
Taylor & Francis Group, an **informa** business

Translated from Russian by V.E. Riecansky

CRC Press
Taylor & Francis Group
6000 Broken Sound Parkway NW, Suite 300
Boca Raton, FL 33487-2742

© 2019 by CISP
CRC Press is an imprint of Taylor & Francis Group, an Informa business

No claim to original U.S. Government works

Printed on acid-free paper

International Standard Book Number-13: 978-0-367-20151-7 (Hardback)

This book contains information obtained from authentic and highly regarded sources. Reasonable efforts have been made to publish reliable data and information, but the author and publisher cannot assume responsibility for the validity of all materials or the consequences of their use. The authors and publishers have attempted to trace the copyright holders of all material reproduced in this publication and apologize to copyright holders if permission to publish in this form has not been obtained. If any copyright material has not been acknowledged please write and let us know so we may rectify in any future reprint.

Except as permitted under U.S. Copyright Law, no part of this book may be reprinted, reproduced, transmitted, or utilized in any form by any electronic, mechanical, or other means, now known or hereafter invented, including photocopying, microfilming, and recording, or in any information storage or retrieval system, without written permission from the publishers.

For permission to photocopy or use material electronically from this work, please access www.copyright.com (http://www.copyright.com/) or contact the Copyright Clearance Center, Inc. (CCC), 222 Rosewood Drive, Danvers, MA 01923, 978-750-8400. CCC is a not-for-profit organization that provides licenses and registration for a variety of users. For organizations that have been granted a photocopy license by the CCC, a separate system of payment has been arranged.

Trademark Notice: Product or corporate names may be trademarks or registered trademarks, and are used only for identification and explanation without intent to infringe.

Visit the Taylor & Francis Web site at
http://www.taylorandfrancis.com

and the CRC Press Web site at
http://www.crcpress.com

Contents

Foreword		**viii**
Introduction		**x**
1.	**Fundamentals of mechanics of strength and plasticity of metals**	**1**
1.1.	Basic concepts, postulates and method in the classical mathematical theory of plasticity (flow theory)	1
1.2.	The defining relations of the theory of plasticity (particular laws of metal deformation)	12
1.2.1.	*The tensor defining relations*	12
1.2.2.	*Scalar defining relations*	24
1.3.	Fundamentals of the classical mathematical theory of creep of metals	27
1.4.	Modern approaches to the development of the mathematical theory of irreversible strains and the formulation of a scientific problem	41
1.4.1.	*Plasticity theory*	41
2.	**Fundamentals of the phenomenological theory of fracture and fracture criteria of metals at high plastic strains**	**57**
2.1.	Basic concepts, assumptions and equations of the phenomenological theory of the fracture of metals	57
2.2.	Criteria of ductile fracture of metals	66
2.3.	Modern approaches to the development of the theory of ductile fracture and the formulation of a scientific problem	69
3.	**Fundamentals of the physics of strength and plasticity of metals**	**79**
3.1.	Basic concepts and assumptions of the dislocation theory of plasticity	79

3.2.	Theoretical description of plastic deformation	98
3.2.1.	*Multilevel character of plastic deformation*	98
3.2.2.	*Structure and properties of metals with developed and intense plastic strains*	108
3.2.3.	*Methods of theoretical description of plastic deformation*	116
3.2.4.	*Physical (microstructural) models of creep of metals*	121
3.3.	Basic concepts and provisions of the physics of fracture of metals	128
4.	**A physico-phenomenological model of the single process of plastic deformation and ductile fracture of metals**	**138**
4.1.	General provisions of the model	138
4.2.	The scalar defining equation of viscoplasticity	145
4.3.	Scalar model of the plasticity of a hardening body (cold deformation of metals)	148
4.4.	Model of ductile fracture of metals	149
4.5.	Obtaining a generalized law of viscoplasticity based on a scalar law	154
5.	**A physico-phenomenological model of plasticity at high cyclic deformation and similar cold deformation**	**160**
5.1.	The experimental basis of the model	160
5.2.	The defining equations of large cyclic deformation and deformation close to it	165
6.	**Physico-phenomenological models of irreversible strains in metals**	**169**
6.1.	Model of evolution of a microstructure under irreversible deformation of metals	169
6.2.	Kinetic physical-phenomenological model of dislocation creep, controlled by thermally activated slip of dislocations	170
6.3.	Kinetic physico-phenomenological model of long-term strength of metals	176
6.3.1.	*General information about long-term strength*	176
6.3.2.	*Model of long-term strength. The general case of loading*	180
6.3.3.	*Modelling of the process of testing samples for long-term strength under conditions of stationary*	

	thermomechanical loading	182
6.4.	Stress relaxation model	183
7.	**Experimental verification of adequacy of models**	**186**
7.1.	Scalar viscoplasticity model	186
7.1.1.	*Methodology for checking the adequacy of the model*	186
7.1.2.	*Results of model verification*	188
7.2.	Model of ductile fracture of metals	196
7.3.	Creep model	198
7.4.	Stress relaxation model	203
7.5.	Model of long-term strength	203
7.6.	Model of evolution of the structure in processes of irreversible deformation of metals	205
7.7.	The model of a large cyclic and near-plastic deformation	208
8.	**Mathematical formulation and examples of solving applied problems of the physico–mathematical theory of plasticity**	**215**
8.1.	Mathematical formulation of problems	215
8.2.	Examples of development, research and improvement of processes of processing of metals by pressure on the basis of mathematical modelling	218

Conclusion 229
References 230
Index 238

Foreword

In the monograph proposed for the first time, an attempt was made to systematically present the scientific results obtained in the study of a single physical process of irreversible deformation and brittle fracture of metals within the framework of the structural phenomenological approach.

The essence of this approach is the integration of macro- and micro-representations, methods of macro- and micro-description of the process. Practical realization of this association took place in the form of a new theory, the exposition of which is the goal of this book.

The theory has a deductive character, that is, it is built on its own postulates. The generalization of the equations of the theory, obtained initially for a uniaxial stressed state, to a volume stress-strain state required further development of the foundations of the classical mathematical theory of plasticity. Two theorems are formulated and proved which are generalizations of the Drucker postulate and the von Mises maximum principle of the classical mathematical theory of plasticity (flow theory) to a viscoplastic medium. Consequently, the theory includes as a special case the classical mathematical theory of plasticity, which follows from the general theory under the assumption that there is no thermodynamic return during deformation.

A special feature of the theory is its construction in finite increments, which made it possible immediately to bypass a number of problems of the classical theory, for example, the problem of dividing large deformations into elastic and irreversible components.

This feature of the theory, together with the presence of evolution equations for the actual structural parameters-the scalar dislocation and microcrack densities-provides the formulation and solution of practical problems in which nonlinear processes of large irreversible deformations occurring under conditions of a nonstationary stress-strain state and a temperature field are considered.

Foreword

The physico-mathematical theory of strength and plasticity for the first time consistently takes into account the continuous change in the structure of materials during deformation and the accumulation of deformation damage.

The novelty of the theory, in our opinion, is that, for the first time, a scheme for describing plastic deformation and viscous destruction, evolution of structure, creep processes, long-term strength of metals and stress relaxation is proposed in the framework of a unified approach and a unified model.

It is not difficult to see that on this basis it is possible to further expand this scheme and include in it a forecasting model for the residual resource, a model for determining the mechanical characteristics of quasi-samples of standard mechanical properties in deformed semi-finished products by the method of mathematical modeling.

To apply the theory to the development and study of processes of irreversible deformation and viscous destruction, it is necessary to create a specialized software product of a new generation for computer engineering analysis. Existing means, for example *DEFORM-3D*, can not in general be used to implement the model of the deformation process formulated within the framework of the new theory. This is due to the fundamental differences in the algorithms for solving the applied problems of the classical mathematical theory of plasticity and the new theory.

Introduction

Among scientific workers, it is considered a bad practice to go beyond the scope of one's research (one's science) with an attempt to 'shift' another scientific discipline. There are good reasons for this. The modern level of development of scientific disciplines, especially fundamental ones, in which mathematical methods of description and research are widely used, is very high. In order to master the results of the long-term development of a particular science, which is a prerequisite for bold attempts to contribute to its further development, it takes a lot of time and labour. At the same time, as a rule, each scientific discipline has its own characteristic way of thinking, connected with its methods.

Therefore, scientists are sometimes skeptical of attempts by their colleagues to contribute to the development of 'not their own' science.

On the other hand, it is well known that scientific research conducted at the junction of two related disciplines is often the most effective and is accompanied by a combination of scientific directions.

The process of combining disciplines, along with their differentiation, is an objective law of the development of science. A vivid example is the creation of statistical physics, which is a synthesis of molecular–kinetic teaching and thermodynamics. This example demonstrates a synthetic approach to the development of a scientific problem – the unification of the phenomenological and statistical approaches to the study and description of phenomena and processes in macroscopic systems.

The fundamental scientific discipline – the mechanics of a deformable solid (MDS) consists of four sections: theories of elasticity, plasticity, creep, fracture mechanics of materials and their applications. At present, MDS is a highly developed science with a powerful mathematical apparatus that allows solving a wide range of applied problems related to the design of various designs and technologies for processing materials. The applied value of the theory in recent years has increased significantly in connection with the intensive development of computational mechanics and computers.

The main applied task of all four sections of the discipline is the calculation of stress fields, strains, strain rates and deformation damage of solids of arbitrary geometric shape loaded with an arbitrary system of external forces. Information about these fields is obtained as a solution of the initial-boundary value problem of mathematical physics.

Since deformation of continuous solids is one of the forms of their mechanical motion, the mathematical formulation of the initial-boundary problem includes differential equations of motion and kinematic relations, which are closed by the defining equations. The latter are models of deformable solids: models of elasticity, plasticity, creep, material with damage. Therefore, the task of MDs is the development of these models – the laws of deformation of solids.

For centuries and now, despite the creation and increasing use of ceramics, plastics, glasses, composites, the main structural material in all branches of engineering industry are metals that have a unique combination of strength, elasticity, ductility and viscosity. Therefore, specialists of MDS pay great attention to the development of models of metallic materials. The subject of the study in this work are metals and their alloys.

MDS, like general mechanics, is a phenomenological science. The first hypothesis in the phenomenological approach to the study of the motion of matter is the hypothesis of a continuous medium, i.e., as a material model, a continuous deformable medium is assumed. However, it is well known that the deformation behaviour of materials (the properties of materials) is determined by their internal structure. In this case, during deformation, the structure changes significantly, and, consequently, at every stage of deformation we are dealing, generally speaking, with a new material.

The failure to take into account the structure of the material and its evolution under deformation is the cause of problems that have been clearly formulated by the second half of the last century, which have held back the development of MDS for years. These problems are associated with the non-linearity of the processes of large irreversible strains, taking into account the history of deformation and the evolution of the structure of the material. In more detail, these problems are formulated in the first and second chapters of the book, on the content and current state of the mathematical theories of plasticity, creep, and ductile fracture of metals. Numerous attempts to solve these problems, undertaken for decades within the framework of a purely phenomenological approach, have so far not yielded meaningful results.

Irreversible deformation and destruction of metals are a subject of investigation of another fundamental discipline – the physics of strength and plasticity, which is a section of the physics of solids. This discipline uses microstructural and statistical approaches to the study of the motion of matter. It uses a discrete atomic model of the material, studies and takes into account the evolution of the structure under deformation. The laws of deformation of materials are established in the form of some model representations, based on the analysis of micromechanisms of processes occurring in the atomic model of a material when it is loaded by external forces.

The physics of strength and plasticity develops uniaxial deformation laws. It lacks methods of obtaining generalized laws necessary for solving practical problems. The desire to construct a more accurate physical deformation model determines the presence in the equations of a large number of parameters, the physical meaning of which often remains unclear. This makes it practically impossible to use physical models to calculate real processes.

The natural direction of the further development of MDS and the physics of strength and plasticity is the unification of micro- and macro-representations about irreversible deformations and fractures, methods of micro- and macro-description of these physical processes. Therefore, in spite of the partial division of thoughts formulated at the beginning of this introduction, the author, with a small group of assistants (undergraduates and graduate students), began systematic research on the development of the formulated direction in the early 1990s. The results obtained over the past years are generalized and presented to the reader in this monograph.

Taking into account the orientation of the book, the author found it necessary to preface a new material – the physical and mathematical theory of irreversible strainss of metals – with an exposition of the elementary foundations of mechanics and the physics of strength and plasticity of metals, believing that this should help mechanical engineers and physicists get acquainted with its content.

Section I

Current State of Mechanics and Physics of Strength and Plasticity of Metals

Fundamentals of mechanics of strength and plasticity of metals

1.1. Basic concepts, postulates and method in the classical mathematical theory of plasticity (flow theory)

Metals and metallic alloys are the main structural materials in engineering. This is due to a rational combination of the characteristics of strength, elasticity, ductility and toughness. The ductility of metals is essential.

The physical process of plastic deformation underlies one of the oldest methods of metalworking – plastic forming of metals. The ability to plastic deformation determines the performance of metal structures, including parts of machines and mechanisms. The scientific basis for the design of metal structures and machine components, as well as the technological processes of plastic shaping (forging and stamping), along with theories of elasticity and creep, is the theory of plasticity.

The applied task of these scientific disciplines is the calculation of the parameters of the above objects, ensuring their optimal or rational functioning. This problem is solved by determining the stresses and strains in the workpiece being machined, the structure, the part.

It is believed that the theory of plasticity originates from the work of Saint-Venant, published in 1871 [1]. The main stages in the formation and development of the theory of plasticity, as well as biographical information about scientists who contributed to the development of the theory, are given in the interesting concluding section 'A Brief Historical Reference on the Chapters' of the widely known textbook on the mechanics of plastic forming of metals by Prof. V.L. Kolmogorov [2].

The object of studying the theory of plasticity as a fundamental scientific discipline is a special form of mechanical motion of deformable solids – plastic deformation and its models. This form of motion is not considered by theoretical mechanics, in which the model of a solid non-deformable body is adopted [3]. The classical mathematical theory of plasticity is, like mechanics in general, a purely phenomenological discipline, has its own axiomatics, that is, the theory is built on the basis of its own postulates (principles). The construction is carried out within the framework of the classical mechanics of Galileo–Newton and the paradigm that is unified with analytical mechanics.

The main concepts are [4]: *physical space, time, mass, force* (primary concepts in mechanics), *region V of physical space with boundary S, continuous medium, elementary volume (material particle); displacement vector, velocity and acceleration of particle motion, kinetic, internal and free energy, entropy and particle temperature; tensors of deformations, strain rates and stresses, loading processes in the stress space $\sigma_{ij}(t)$ and deformation in the strain space $\varepsilon_{ij}(t)$, where σ_{ij} is the stress tensor, ε_{ij} is the strain tensor, and t is the time.*

Thermodynamic concepts and relations are introduced as applied to a particle of a continuous medium. The particle is identified with a mathematical point.

The loading process is called simple if the stress state of the particle changes over time in such a way that the end of the stress vector in the six-dimensional image space of the symmetrical stress tensor moves along the ray emanating from the origin, i.e., the straight line is the path of the stress vector. With other trajectories (curvilinear, broken) – the loading is complex. Cyclic loading is

Fundamentals of Mechanics of Strength and Plasticity

characterized by a periodic change in the sign of the stress, can be simple and complex.

Definitions of the basic concepts are given *a priori*. Therefore, the relations between their quantitative measures (mathematical objects) derived on the basis of certain initial postulates can acquire the meaning of substantive (physical) laws only under the condition of experimental justification of the derivations derived or consequences arising from them.

Initial postulates: *macroscopic continuity, homogeneity and isotropy of the deformed body, homogeneity of stress and strain states in an elementary (representative) volume ΔV (material particle)* in which all the vector and scalar quantities characterizing the thermomechanical state are average (integral) in the sense of the averages in statistical physics, *macrophysical definability* [5], *fluidity conditions, loading functions, a single curve*.

The basic applied problem of the theory of plasticity by definition of the stress–strain state of a body, which in general has an arbitrary shape and is loaded with an arbitrary system of external forces, is solved by setting the *initial-boundary value problem of plasticity theory* [2]. It includes the following equations.

1. Equilibrium equations, into which the equations of motion of a continuous medium (Newton's second law) are transformed, if mass forces are neglected:

$$\sigma_{ij,j} = 0^{1)} \quad i,j = x, y, z. \tag{1.1}$$

2. Geometric Cauchy relations for small strains:

$$\varepsilon_{ij} = \frac{1}{2}(u_{i,j} + u_{j,i}), \tag{1.2}$$

where $\varepsilon_{ij} = \varepsilon_{ij}^e + \varepsilon_{ij}^p$, ε_{ij}^e, ε_{ij}^p are the elastic and plastic components of the strain tensor; u_i are the projections onto the coordinate axes of the displacement vector of the particle. The strains must satisfy the compatibility conditions [2]

$$\frac{\partial^2 \varepsilon_{ij}}{\partial x_{kl}} + \frac{\partial^2 \varepsilon_{kl}}{\partial x_i \partial x_j} = \frac{\partial^2 \varepsilon_{il}}{\partial x_k \partial x_j} + \frac{\partial^2 \varepsilon_{kj}}{\partial x_i \partial x_l}.$$

3. The equation of heat conductivity

$$c\rho \frac{dT}{dt} = (\lambda T_{,i})_{,i} + \sigma_{ij}\dot{\varepsilon}_{ij}, \qquad (1.3)$$

where c, λ are the coefficients of heat capacity and thermal conductivity of the material being deformed; ρ is its density; T is the thermodynamic temperature; $\dot{\varepsilon}_{ij} = d\varepsilon_{ij}/dt$ is the plastic strain rate tensor.

The paper deals with the theory of plastic deformation of compact metallic materials. If the incompressibility condition $\varepsilon_x + \varepsilon_y + \varepsilon_z = 0$ and $\rho = $ const is adopted, then the strain tensors and strain rates coincide with their deviators.

When the boundary-value problem is formulated in the velocities, equations (1.2) are replaced by the Saint-Venant geometric relations for the strain rates

$$\dot{\varepsilon}_{ij} = \frac{1}{2}(v_{i,j} + v_{j,i}), \qquad (1.4)$$

where v_i are the projections onto the coordinate axes of the velocity vector of the material particle.

The system of ten differential equations (1.1)–(1.3) contains 16 unknown (sought) functions of coordinates and time: $\sigma_{ij}(x_i, t)$, $\varepsilon_{ij}(x_i, t)$, $T(x_i, t)$, $u_i(x_i, t)$. It is closed by the *defining equations* describing the relationship of stresses to strains (or strain rates) and the mechanical properties of the material. To obtain a particular solution, the system is supplemented by boundary conditions.

The defining relations are a mathematical model of the plastic deformation of a material (the law of deformation). Therefore, one of the main tasks of the mathematical theory of plasticity as a fundamental discipline is the establishment of laws of deformation of materials in various thermomechanical conditions and structural states.

The most general formulation of the laws of deformation is given on the basis of certain extreme principles [6–8]. It is assumed that the loading process (the change in the state of the elementary volume of the deformed medium) can be described by a finite set of pairs of parameters σ_{ij}, T, ε_{ij}^p, q, k, where q is the hardening parameter, which is usually taken to be the plastic strain, referred to the unit volume, $q = A = \int \sigma_{ij} d\varepsilon_{ij}^p$, or the plastic deformation intensity accumulated by the particle $q = \int d\varepsilon^p$ ($d\varepsilon^p$ is the intensity of the increment of plastic

Fundamentals of Mechanics of Strength and Plasticity 5

deformations); k is the parameter associated with the yield strength of the material. Time implicitly enters through q.

Since the temperature T is a scalar quantity, it is excluded from the number of determining parameters and is taken into account by the dependence $k(T)$, while, as a rule, an isothermal deformation process is considered. A mixed problem *can also be posed, including the mechanical and thermal problems*. The basis for the formulation and solution of the second problem is equation (1.3).

It is postulated that the current (actual) thermomechanical state of the elementary volume of a body under loading by its system of external forces can be described in a six-dimensional stress space by a certain function of the determining parameters

$$f(\sigma_{ij}, \varepsilon_{ij}^p, q, k) = 0, \tag{1.5}$$

which is called the *loading function*. It is also assumed that for small strains, the total strains can be represented as the sum of reversible (elastic) and irreversible (plastic) components. For fixed parameters, the function (1.5) describes a hypersurface in the six-dimensional space of a symmetric bivalent stress tensor σ_{ij} (stress space). If the conditions for the commencement of the Huber–von Mises fluidity and the independence of f from the first and third invariants of the stress tensor are accepted, then the loading function takes a particular and specific form

$$f = \frac{3}{2} s_{ij} s_{ij} - \sigma_T^2 = 0, \tag{1.6}$$

where σ_T is the yield strength of the isotropic material.

The surface described by (1.6) is called the *surface of the beginning of plasticity*, when $\varepsilon_{ij}^p = 0$, $q = 0$. In the case of an elastoplastic material, it separates the region of elastic stress states, where the stresses and strains are related by Hooke's law (inside the area bounded by the surface) from the stress states at which the elementary volume passes into a plastic state (the stressed states are on the surface). Consequently, in the elastic region $f < 0$. For a rigid-plastic body in the inner region, bounded by the hypersurface of plasticity (1.6), the material is absolutely rigid. When the hardening material is deformed in (1.6), instead of σ_T, according to the Huber–von Mises plasticity condition, the current stress intensity σ.

With hardening σ, the surface of plasticity, which in this case is called the *loading surface*, also increases and changes. The change in the loading surface describes the hardening of the material to be deformed. In the general case of deformation of a material, it can change the shape and position in the space σ_{ij}. In an ideal plastic material (σ_T = const), the loading surface does not change (fixed) during deformation and is called the *yield surface*.

The following properties of the loading surface are postulated: it is closed, but in some directions it can extend to infinity; does not pass through the origin; any ray drawn from the origin crosses it only once, that is, it does not have concave sections.

The loading surface, or part of it, at each point of which there is a single outward normal and, therefore, it is differentiable with respect to σ_{ij}, is said to be *smooth*, and the loading points on it are *regular*.

The concepts of *loading*, *unloading* and *neutral loading* for regular points are introduced. If the stress state of the material particle described by the six-dimensional vector σ_{ij} belongs to the loading surface Σ (hence, the plasticity condition (1.6) is satisfied) and the loading vector $d\sigma_{ij}$ is directed outwards, this process is accompanied by an increase in the plastic deformation of the particle $d\varepsilon_{ij}^p > 0$, the change of the loading surface (in the case of the hardening body it assumes the position Σ′) and is called the *loading* (Fig. 1.1). After loading, the stressed state of the particle is described by the vector σ'_{ij} (shown by the dashed line).

With the direction of the vector $d\sigma_{ij}$ inside Σ, the resulting stress state (point A in Fig. 1.1) is in the elastic region. In this ij case, $d\varepsilon_{ij}^p = 0$, the stresses and strains are related by the generalized Hooke's law, and we are dealing with the *unloading process*. The loading process is called *neutral* if $d\sigma_{ij}$ is directed along the tangent

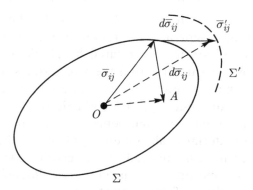

Fig. 1.1. Definition of the loading and unloading processes of an elastoplastic hardening body.

Fundamentals of Mechanics of Strength and Plasticity

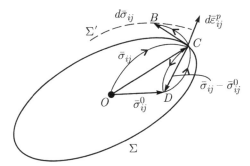

Fig. 1.2. Postulated loading and unloading cycle of the hardening elastoplastic body, on the basis of which the principle of maximum work of plastic deformation is formulated.

to the surface Σ. In the case of a hardening body $d\varepsilon_{ij}^p = 0$ and stresses are related with the strains also related by Hooke's law.

The basis for constructing the plasticity model of a hardening body is the Drucker postulate (Drucker, D.C.) [6], which states that the *work of additional stresses on the increments of plastic deformations caused by them during the loading and unloading cycle is positive. In the stress space* we consider the cycle shown in Fig. 1.2. The material particle from the original natural state, in which there are no stresses and deformations in the particle (the point O is the origin), is loaded along a certain path to the yield point (point C) and unloaded to point D. This state is accepted in the argument for the original. In this state, the particle is loaded to the point C, is loaded with the vector $d\boldsymbol{\sigma}_{ij}$ (point B) and unloaded to the initial state (point D). In this case, the loading and unloading paths are arbitrary and lie in the elastic region except for the addition loading by the vector $d\sigma_{ij}$.

From the considered cycle, taking into account the fact that the work of additional stresses on reversible elastic strains under conditions of a closed deformation path is zero, according to the postulate it follows that

$$\left[\left(\sigma_{ij}-\sigma_{ij}^0\right)+d\sigma_{ij}\right]d\varepsilon_{ij}^p > 0. \qquad (1.7)$$

If we take the point O as the initial and final states, that is, $\sigma_{ij}^0 = 0$ then (1.7) takes the form

8 *Physico-Mathematical Theory of High Irreversible Strains*

$$(\sigma_{ij} + d\sigma_{ij}) d\varepsilon_{ij}^p > 0. \tag{1.7a}$$

If the initial state is taken as the state at the yield point (point C in Fig. 1.2), when $\sigma_{ij}^0 = \sigma_{ij}$, then it follows from (1.7) that

$$d\sigma_{ij} d\varepsilon_{ij}^p > 0. \tag{1.8}$$

This expression is considered as a condition for stable deformation beyond the elastic limit of a hardening elastoplastic body in the general case for a volume stress–strain state. We note at once that inequality (1.8) imposes a restriction on the possibility of describing plastic deformation. It excludes from consideration materials with an incident (non-monotonic) deformation diagram when $d\sigma/d\varepsilon^p < 0$. Therefore, the description of the deformation of materials with a non-monotonic diagram is one of the problems of the classical theory.

If $\sigma_{ij}^0 \neq \sigma_{ij}$, then the difference $\sigma_{ij} - \sigma_{ij}^0$ can be arbitrarily greater than $d\sigma_{ij}$, and then

$$(\sigma_{ij} - \sigma_{ij}^0) d\varepsilon_{ij}^p > 0. \tag{1.9}$$

From this inequality follows the *principle of the maximum work of plastic deformation* for a hardening plastic body:

$$\sigma_{ij} d\varepsilon_{ij}^p > \sigma_{ij}^0 d\varepsilon_{ij}^p. \tag{1.10}$$

For any given value of the components of the plastic deformation increment, the increment of the plastic deformation work $\sigma_{ij} d\varepsilon_{ij}^p$ has a maximum value for the actual stress state σ_{ij} in comparison with all possible stress states σ_{ij}^0 satisfying the condition $f(\sigma_{ij}) < 0$.

Like other differential principles of mechanics, the principle of the maximum work of plastic deformation allows in this case to separate the true stress states from all possible ones in deformation of the body.

The increment in the work of plastic deformation is a function of the stress components. The latter are not independent arguments, since during plastic deformation they must simultaneously satisfy the plasticity condition (1.6). Therefore, the maximum of the function of increment of the work of plastic deformation, declared by the above-stated principle, is conditional. In mathematics, several methods for solving problems on the conditional extremum of a function of several variables have been developed. In the theory of plasticity, the condition for the maximum of the function $dA = \sigma_{ij} d\varepsilon_{ij}^p$, in the presence of the coupling equation $f(\sigma_{ij}) = 0$ (plasticity condition), is written using the Lagrange multiplier method [9] as

$$\frac{\partial}{\partial \sigma_{ij}} \left(\sigma_{ij} d\varepsilon_{ij}^p - d\lambda f \right) = 0, \qquad (1.11)$$

where $d\lambda$ is the Lagrange multiplier.

After differentiation, an equation is obtained which is one of the vertices of the mathematical theory of plasticity and is called the *associated flow law (with the plasticity condition)*:

$$d\varepsilon_{ij}^p = d\lambda \frac{\partial f}{\partial \sigma_{ij}}. \qquad (1.12)$$

It follows from (1.12), firstly, that the loading function f is a *plastic potential* in the stress space and, secondly, that the vector $d\varepsilon_{ij}^p$ is directed along the normal to the loading surface, since $\frac{\partial f}{\partial \sigma_{ij}} = \mathbf{n}$ (**n** is is the unit vector of the normal to the surface f), and $d\lambda$ is a scalar.

It is proved that if the first invariant of the stress tensor is included explicitly or implicitly as arguments in f, then the plastic deformation proceeding according to the law (1.12) satisfies the incompressibility condition $d\varepsilon_{ii}^p = 3 d\varepsilon_0 = 0$, that is, the tensor $d\varepsilon_{ij}^p$ coincides with its deviator.

Substitution of the components of the increment of plastic deformations (1.12) into the expression for the intensity of the increment of plastic strains

$$d\varepsilon^p = \left(2/3 \, d\varepsilon_{ij}^p d\varepsilon_{ij}^p \right)^{1/2}$$

the value of $d\lambda$ is defined as

$$d\lambda = \sqrt{\frac{3}{2}} d\bar{\varepsilon}^p \bigg/ \left(\frac{\partial f}{\partial \sigma_{ij}} \cdot \frac{\partial f}{\partial \sigma_{ij}} \right)^{\frac{1}{2}}. \qquad (1.13)$$

Taking into account the decomposition of the stress tensor into a deviator and spherical tensors:

$$\sigma_{ij} = s_{ij} + \delta_{ij}\left(\frac{1}{3}\delta_{kl}\sigma_{kl}\right),$$

$$s_{ij} = \sigma_{ij} - \delta_{ij}\left(\frac{1}{3}\delta_{kl}\sigma_{kl}\right),$$

the derivative of the loading function with respect to the components of the tensor σ_{ij} is replaced by the derivative with respect to the components of the stress deviator s_{ij}:

$$\frac{\partial f}{\partial \sigma_{ij}} = \frac{\partial f}{\partial s_{ij}}\frac{\partial s_{ij}}{\partial \sigma_{ij}} = \frac{\partial f}{\partial s_{ij}} - \delta_{ij}\left(\frac{1}{3}\delta_{kl}\frac{\partial f}{\partial s_{kl}}\right), \quad (1.14)$$

where δ_{ij} is the Kronecker symbol ($\delta_{ij} = 1$ for $i = j$ and $\delta_{ij} = 0$ for $i \neq j$).

The equation of the associated flow law (1.12) takes the form

$$d\varepsilon_{ij}^p = d\lambda\left(\frac{\partial f}{\partial s_{ij}} - \frac{1}{3}\delta_{ij}\delta_{kl}\frac{\partial f}{\partial s_{kl}}\right). \quad (1.15)$$

If a certain loading function f is adopted for a particular material, then (1.15) are the equations (law) of its plastic deformation in the general case of a volume stress–strain state.

Instead of the principle (1.10), for the formulation of the plasticity model of a hardening body in stress space one can also use the principle of the maximum dissipation rate of mechanical work $D = \sigma_{ij} \cdot \dot{\varepsilon}_{ij}^p$. *For fixed parameters* ε_{ij}^p, q *of the loading function f, for any given value of the velocity components of the deformation* $\dot{\varepsilon}_{ij}^p$, *we have the inequality*

$$\sigma_{ij}\dot{\varepsilon}_{ij}^p > \sigma_{ij}^0\dot{\varepsilon}_{ij}^p, \quad (1.16)$$

where σ_{ij} are the real values of the stress components corresponding to a given value $\dot{\varepsilon}_{ij}^p$; σ_{ij}^0 are the components of any possible stress state allowed by the given loading function $f\left(\sigma_{ij}^0, \varepsilon_{ij}^p, q, k\right) < 0$.

This is the *von Mises maximum principle*. It can be seen that (1.16) can be obtained by differentiating (1.10) with respect to time at constant stresses. In this case, the associated flow law will be written as

Fundamentals of Mechanics of Strength and Plasticity 11

$$\dot{\varepsilon}_{ij}^p = \mu^0 \frac{\partial f}{\partial \sigma_{ij}}, \qquad (1.17)$$

where $\mu^0 = \sqrt{\dfrac{\dot{\varepsilon}_{hk}^p \dot{\varepsilon}_{hk}^p}{\partial f/\partial \sigma_{mn} \partial f/\partial \sigma_{mn}}} = \dfrac{d\lambda}{dt}$.

The construction of a model of a hardening plastic body can also be based on the definition of a *dissipative function* [6]

$$D = \sigma_{ij}\dot{\varepsilon}_{ij} = D(\dot{\varepsilon}_{ij}, \varepsilon_{ij}, q, k), \qquad (1.18)$$

which in the six-dimensional space of the symmetric strain rate tensor for fixed parameters ε_{ij}, q and k describes the *surface of an equal level of dissipation of mechanical work per unit volume per unit time*. In this case, for fixed parameters ε_{ij}, q, along with the real values $\dot{\varepsilon}_{ij}$, we introduce into consideration possible $\dot{\varepsilon}_{ij}^0$, for which

$$D(\dot{\varepsilon}_{ij}^0, \varepsilon_{ij}, q, k) \leq D(\dot{\varepsilon}_{ij}, \varepsilon_{ij}, q, k). \qquad (1.19)$$

Similarly to the von Mises maximum principle (1.16), *we formulate the principle of the maximum dissipation rate* in the strain rate space – the *Ziegler principle*:

$$\sigma_{ij}\dot{\varepsilon}_{ij} \geq \sigma_{ij}\dot{\varepsilon}_{ij}^0. \qquad (1.20)$$

The associated flow law in this case has the form

$$\sigma_{ij} = \lambda \frac{\partial D}{\partial \dot{\varepsilon}_{ij}}, \quad \lambda = D \bigg/ \left(\frac{\partial D}{\partial \dot{\varepsilon}_{ij}} \dot{\varepsilon}_{ij}\right). \qquad (1.21)$$

Specific material models, as in the case of determining the loading function f, are determined by the assumption of the structure of the function D. Using the defining relations of the form (1.21), the boundary problem of plasticity is posed and solved at velocities.

Thus, the plastic body model is introduced in two ways: either through the definition of the loading function f, or through the definition of the dissipative function D. In both cases, the corresponding extreme principles – the principles of maximum – are formulated.

It is important to note that the indicated paths are equivalent, since the definition of the function D is possible if the model of the plastic body is given by the relations (1.5), (1.17) [8].

In conclusion of a brief review of the basics of the mathematical theory of plasticity, let us draw the reader's attention to the following fact which has an epistemological significance. There are clear parallels between the method of constructing Lagrange's analytical mechanics, which deals with the study of general laws of motion of systems of solids, and the method of plastic deformation mechanics (the mathematical theory of plasticity), which deals with the study of general laws of deformation of deformable bodies: the Lagrange function is a loading function; the principle of least action, for example Hamilton – the principle of maximum work of plastic deformation; the Lagrange motion equations are the equations of the associated flow law. These parallels clearly indicate a certain unity of the methodology of mechanics as a whole as a phenomenological science having a deductive character.

The material presented is the basis of one direction of the general mathematical theory of plasticity, called the *flow theory*. A second trend is developing – *the theory of processes* [10], the founder of which is one of the most outstanding mechanics scientists of the 20th century, A.A. Il'yushin [5]. The main difference between the theory of processes and the flow theory lies in the method of geometric interpretation of the deformation process and in the method of constructing the defining relations. The account and analysis of the theory of processes is not included in the tasks of this book.

Particular models of the plasticity theory are a consequence of different formulations of the plasticity condition and, respectively, the loading functions.

Let us consider in retrospect some of these models that have great practical importance.

1.2. The defining relations of the theory of plasticity (particular laws of metal deformation)

1.2.1. The tensor defining relations

Widely used in calculations and mathematical modelling, including technological shaping operations of pressure metal working (PMW), are the defining relationships of the *theory of plasticity of isotropic material with isotropic hardening*.

Fundamentals of Mechanics of Strength and Plasticity

In this theory, the Huber–von Mises plasticity condition and, accordingly, the loading function are taken in the form

$$f(\sigma_{ij}) = \frac{3}{2} s_{ij} s_{ij} - [\Phi(q)]^2 = 0, \qquad (1.22)$$

where q is the Udquist hardening parameter, $q = \int d\varepsilon^p$ (integration is carried out along the strain path).

The differentiation of the function (1.22) with respect to the components s_{ij} and multiplication by δ_{ij} leads to the result

$$\delta_{ij} \frac{\partial f}{\partial s_{ij}} = 3 \delta_{ij} s_{ij} = 0.$$

Then, substituting (1.22) into the associated flow law (1.15), we obtain

$$d\varepsilon_{ij}^p = d\lambda \frac{\partial f}{\partial s_{ij}} = 3 d\lambda s_{ij}. \qquad (1.23)$$

A comparison of (1.23) with (1.12) yields

$$\frac{\partial f}{\partial \sigma_{ij}} = \frac{\partial f}{\partial s_{ij}} = 3 s_{ij}. \qquad (1.24)$$

Thus, in the case of a loading function of the form (1.22), (1.24) holds and, according to (1.13), the Lagrange multiplier is defined as

$$d\lambda = \frac{1}{2} \frac{d\varepsilon^p}{\sigma}, \qquad (1.25)$$

where $\sigma = \sqrt{\frac{3}{2} s_{ij} s_{ij}}$ is the stress intensity.

The substitution of (1.24) and (1.25) into the equation of the associated flow law (1.12) gives the defining equations of the isotropic flow theory, having the form

$$d\varepsilon_{ij}^p = \frac{3}{2} \frac{d\varepsilon^p}{\sigma} s_{ij}. \qquad (1.26)$$

Characteristics of the mechanical properties of a particular material enter the flow law (1.26) by means of the ratio $d\varepsilon^p/\sigma$, which, in fact, is a phenomenological coefficient in the phenomenological equation

14 Physico-Mathematical Theory of High Irreversible Strains

(1.26). Its definition is based on the following chain of reasoning. From (1.22) it follows that

$$\sigma = \Phi\left(\int d\varepsilon^p\right) = \Phi(\varepsilon^p), \tag{1.27}$$

i.e., for materials with the yield condition (1.22), the stress intensity is a single-valued function of the strain intensity and does not depend on the form of the stressed state (the hypothesis of a single curve).

For uniaxial tension and compression

$$\sigma = \sigma_1, \quad \int d\varepsilon^p = \varepsilon^p \text{ and } \sigma = \Phi(\varepsilon^p).$$

Therefore, the ratio $d\varepsilon^p/\sigma$ can be determined from the true tensile or compression diagrams. In this case it is convenient to use diagrams in which the elastic region is excluded because of the smallness of the elastic strains in pressure metal working as compared with the plastic deformations. Such diagrams in the coordinates σ–ε are called *deformation diagrams*. Therefore, *the tensor deformation law (1.26) must necessarily be supplemented by the scalar law $\sigma = \Phi(\varepsilon)$* describing the relationship between the stress intensity and the strain rate under a linear stress state.

Equation (1.22) in a three-dimensional space of principal stress σ_j, $j = 1, 2, 3$, describes a circular cylindrical surface of unlimited length, whose axis is equally inclined to the coordinate axes, and the radius is $\sqrt{2/3}\sigma$. The axis of this *Huber–von Mises plastic cylinder* is perpendicular to the deviator plane, whose equation has the form

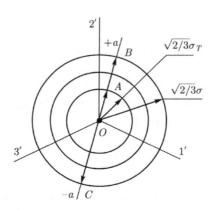

Fig. 1.3. Traces of plasticity cylinders on the deviator plane.

Fundamentals of Mechanics of Strength and Plasticity 15

$$\sigma_1 + \sigma_2 + \sigma_3 = 0,$$

and any vector lying in it characterizes the deviator of some stress state. The line of intersection of the plastic cylinder with the deviator plane is the circle (Fig. 1.3). In the case of a hardening plastic body with increasing plastic strain intensity ε^p, in accordance with (1.22), the stress intensity $\sigma = \sqrt{\frac{3}{2} s_{ij} s_{ij}}$ increases. In this case, the radius of the plasticity cylinder is isotropic (in all directions equally) increased (Fig. 1.3).

The theory of plasticity is called *the flow theory with isotropic hardening*. Taking into account the elastic strainss obeying the generalized Hooke's law, and the additivity of the stresses, its equations have the form

$$d\varepsilon_{ij} = \frac{1}{2G}\left(d\sigma_{ij} - \delta_{ij}\frac{3v}{1+v}d\sigma_0\right) + \frac{3}{2}\frac{d\varepsilon}{\sigma}\left(\sigma_{ij} - \delta_{ij}\sigma_0\right), \qquad (1.28)$$

where

$$d\varepsilon_{ij} = d\varepsilon_{ij}^e + d\varepsilon_{ij}^p;$$

G is the shear modulus; v is Poisson's ratio.

In honor of the authors who first proposed them, they are called the Prandtl–Reuss equations [11, 12].

When describing the plastic deformation in the technological forming operations of pressure metal working, the elastic components of the deformation, as a result of their smallness in comparison with the plastic ones, are usually neglected. By dividing (1.28) by dt we obtain a variant of the equations of the flow theory in velocities

$$\dot{\varepsilon}_{ij} = \frac{3}{2}\frac{\dot{\varepsilon}}{\sigma}s_{ij}, \qquad (1.29)$$

where $\dot{\varepsilon} = d\varepsilon/dt$ is the strain rate intensity.

These equations were obtained by Saint-Venant [13, 14], Levi [15], von Mises [16] and are called the *Saint-Venant–Levi–von Mises equations*.

Using the tensor deformation law in the form (1.29), for the investigation of technological operations and pressure metal working processes, the geometric relations (1.4) enter into the boundary value problem, and for the material being processed, the $\sigma(\dot{\varepsilon})$ dependence, like $\sigma(\varepsilon)$, is determined as a result of specially set experiments on tensile or compression of samples, and which, like $\sigma(\varepsilon)$, is assumed to be invariant under the hypothesis of a single curve with respect to the stress state scheme.

The equations of flow theory (1.28) and (1.29) have found wide application in calculations and studies of technological processes of plastic forming of metals under different thermomechanical conditions. They are the basis of the calculated finite elemental cores of all currently available on the world market domestic and foreign computer software products, designed to address the above tasks. These are the software products: COSMOS, ANSYS, DEFORM, ANTARES, MARC/Autoforge, Forge 2/3, Form-2D, QFORM, RAPID, STAMP, etc., called CAE programs[1].

The deformation laws (1.26) and (1.29), which are valid under the condition of small strains (as will be discussed in detail below), do not take into account the influence on the strain resistance of the load history associated with the scalar properties of metals, i.e., $\dot{\varepsilon}(t)$ and $T(t)$. Accounting for this effect is especially important in the analysis of processes of large hot deformation and, especially, superplastic flow, when metals show the viscosity property most clearly (σ versus $\dot{\varepsilon}$ and T).

The influence of the load history associated with the $\dot{\varepsilon}(t)$ dependence is illustrated in Fig. 1.4. The dependences $\sigma(\varepsilon)$ were constructed from the results of a test of cylindrical samples of 10 kp steel with uposetting at $T = 750°C$ (1023 K) with a step change $\dot{\varepsilon}$ in the following modes: $\dot{\varepsilon} = 10^{-2}$ s^{-1} (curve 1); to $\varepsilon = 0.4$ with $\dot{\varepsilon} = 10^{-3}$ s^{-1} then with $\dot{\varepsilon} = 10^{-2}$ s^{-1} (curve 2); up to $\varepsilon = 0.4$ with

---

1) CAE – Computer-Aided Engineering – computer systems for engineering analysis.

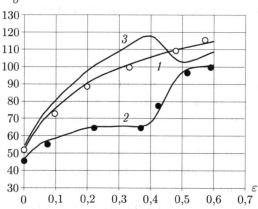

Fig. 1.4. Explanations in the text.

$\dot{\varepsilon} = 1.3 \cdot 10^{-2}$ s^{-1}, then with $\dot{\varepsilon} = 10^{-2}$ s^{-1} (curve 3). It is seen that σ depends not only on the final value of $\dot{\varepsilon}$, but also on the law $\dot{\varepsilon}(t)$ at the previous instants of time. This is a well-known specificity of irreversible processes.

The foregoing deformation laws do not take into account also the influence on the deformation resistance and the load history associated with the vector properties of metals, i.e., with the dependence of σ on the direction in the material. The deformation laws (1.26) and (1.29) take into account isotropic hardening and do not describe anisotropic hardening. The latter is always observed with the directional cold deformation of metals and is manifested, in particular, in the form of the Bauschinger effect. The fact that it is impossible to describe either the Bauschinger effect or the anisotropic hardening using the model (1.26) is clearly demonstrated in Fig. 1.3 by the loading circles.

The elastoplastic hardening body is loaded from the initial natural state (point O), for example, in the direction of the axis (+a) under simple loading conditions. Up to point A the body is in an elastic state. At the point A $\sigma = \sigma_T$, the body changes to a plastic state. With further loading, plastic strains increase and σ increases (the loading surface expands). If the loading stops at point B and continues in the opposite direction (–a), that is, the signs of stress and strain are reversed, then the body will be up to point C in the elastic state. At point C, plastic strains begin at the same (in modulus) stresses, at which the loading process was stopped in the original direction (point B). The model (1.26) describes isotropic hardening.

If the metal in the initial state was isotropic, then as a result of directional plastic deformation (simple loading, monotonic deformation) it becomes anisotropic. Anisotropy is manifested in the dependence of the characteristics of the mechanical properties, including the yield point, on the direction in space. It is said that a deformation anisotropy arises that is visually observed when the isotropic cylindrical specimen is hardened and in the initial state is deformed with a change in the sign of deformation and stress. If the sample is deformed by stretching to some strain ε_1^p and the stress σ_1 is reached and then the tension is changed to compression, then under compression, plastic strains occur at a lower σ_T, i.e., $\sigma_T < \sigma_1$. This phenomenon is known in the literature as the Bauschinger effect.

The phenomenon of deformation anisotropy indicates that the hypothesis of a single curve is not fulfilled for hardening materials and that it is necessary to take it into account when deforming metals

18   *Physico-Mathematical Theory of High Irreversible Strains*

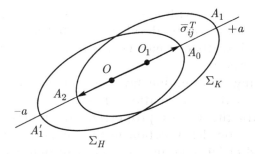

Fig. 1.5. A rigid displacement of the plasticity surface in the space of complete (acting) stresses.

under the conditions of complex loading and deformation, when the stresses and strains in the microvolumes of the deformed material change directions. The appearance of anisotropy is due to structural changes during deformation, since the mechanical properties are derived from the structure.

In order to take anisotropic hardening into account, A.Yu. Ishlinskii and W. Prager assumed that in the process of deformation of an ideal-plastic body, the plasticity surface is rigidly (as a whole) displaced in the stress space. This displacement can be taken into account if the Huber–von Mises plasticity condition is taken in the form

$$f(\sigma_{ij}) = \frac{3}{2}(s_{ij} - a_{ij})(s_{ij} - a_{ij}) - \sigma_T^2 = 0, \qquad (1.30)$$

where a_{ij} is a deviator that varies during deformation by some law b and describes the current coordinates of the centre of the plasticity surface. It is called the deviator of *additional stress*, and the difference $(s_{ij} - a_{ij})$ is the *deviator of the active stress*. Then s_{ij} is the deviator *of the acting (total) stress*. Figure 1.5 shows the initial position of the hypersurface of plasticity Σ_H in the stress space σ_{ij} with the origin at the point O (the initial position of the center of the plasticity surface from which the loading starts). When loaded in the positive direction of the a axis and the plasticity condition is satisfied, $|\sigma_{ij}| = |\sigma_{ij}^T| = |\overline{OA_0}|$, at which $\sigma = \sqrt{\frac{3}{2} s_{ij}^T s_{ij}^T} = \sigma_T$, at first plastic strains begin to grow, and the plasticity surface begins to move rigidly in the direction of loading. After the termination of loading, the current position of the loading surface corresponds to Σ_K, and its centre is at point O_1. The diagram clearly shows that

Fundamentals of Mechanics of Strength and Plasticity 19

in the +a direction in the space of the total stress, the hardening $|\overline{OA_1}| > |\overline{\sigma}_{ij}^T| = |\overline{OA_0}|$ takes place. This hardening is called *translational* or *kinematic*. In this case, in the active stress space, $\sigma = \sigma_T = $ const $\left(|\overline{O_1A_1}| = |\overline{\sigma}_{ij}^T|, \text{ so } |\overline{OA_0}| = |\overline{O_1A_1}|\right)$. Softening $|\overline{OA_2}| < |\overline{\sigma}_{ij}^T| = |\overline{OA_1'}|$ takes please in the reversed direction $-a$. Therefore, after interrupting loading in the initial direction and loading in the reverse direction the material goes to the plastic state at the stress $|\overline{OA_2}| < |\overline{OA_1'}|$, that is, the introduction of translational hardening makes it possible to describe the Bauschinger effect.

A.Yu. Ishlinsky [17] and W. Prager [18] proposed for a_{ij} the simplest linear dependence

$$a_{ij} = c\varepsilon_{ij}^p, \tag{1.31}$$

where c is the constant for a particular material. The model (1.30), (1.31) only qualitatively describes the anisotropic hardening.

Later Yu.I. Kadashevich and V.V. Novozhilov proposed to combine isotropic and kinematic hardening by using the loading function (plasticity condition) of the form

$$f(\sigma_{ij}) = \frac{3}{2}(s_{ij} - a_{ij})(s_{ij} - a_{ij}) - [\Phi(q)]^2 = 0, \tag{1.32}$$

where, unlike (1.31), $a_{ij} = g(\varepsilon^p)\varepsilon_{ij}^p$, and $g(\varepsilon^p)$ is the scalar function of the accumulated intensity of plastic deformation [19]. In this case, $\Phi(q)$ describes isotropic hardening, and a_{ij} kinematic hardening.

In accordance with the associated flow law (1.23), we obtain

$$d\varepsilon_{ij}^p = d\lambda \frac{\partial f(\sigma_{ij})}{\partial (s_{ij} - a_{ij})} = 3d\lambda (s_{ij} - a_{ij}). \tag{1.33}$$

From equation (1.13), taking into account (1.22), we determine $d\lambda$ in the form

$$d\lambda = \sqrt{\frac{3}{2}} \frac{d\overline{\varepsilon}^p}{\left[\frac{\partial f}{\partial (s_{ij} - a_{ij})} \cdot \frac{\partial f}{\partial (s_{ij} - a_{ij})}\right]^{1/2}} = \frac{1}{2} \frac{d\overline{\varepsilon}_{ij}^p}{\Phi(\varepsilon^p)}. \tag{1.34}$$

Substitution of (1.34) into (1.33) gives the defining relations of the (initially) isotropic material with anisotropic hardening (isotropic-translational hardening) of the form

$$d\varepsilon_{ij}^p = \frac{3}{2}\frac{d\bar{\varepsilon}^p}{\Phi(\varepsilon^p)}\left[s_{ij} - g(\varepsilon^p)\varepsilon_{ij}^p\right].\qquad(1.35)$$

From (1.32) for uniaxial tension and subsequent compression, it follows that

$$\begin{aligned}\sigma_t &= \Phi(\varepsilon^p) + \frac{3}{2}g(\varepsilon^p)\varepsilon^p,\\ \sigma_{comp}^T &= -\Phi(\varepsilon^p) + \frac{3}{2}g(\varepsilon^p)\varepsilon^p,\end{aligned}\qquad(1.36)$$

where σ_t and σ_{comp}^T is the stress of the flow upon stretching to different ε^p and the yield strength for subsequent compression, respectively.

Scalar functions of plastic deformation, expressed from (1.36) in the form

$$\Phi(\varepsilon^p) = \frac{\sigma^p - \sigma_{comp}^T}{2},\quad g(\varepsilon^p) = \frac{\sigma^p + \sigma_{comp}^T}{3\varepsilon^p},\qquad(1.37)$$

are determined experimentally when the samples are stretched to different ε^p and σ^p and then compressed to the appearance of plastic deformations for σ_{comp}^T. In this case, the scalar function of the plastic deformation intensity $\Phi(\varepsilon^p)$ is called the *stress function*.

The Kadashevich–Novozhilov model does not describe anisotropic hardening under cyclic loading. When cyclic deformation along a closed trajectory at the beginning and at the end of each cycle in accordance with the model $\varepsilon_{ij}^p = 0$ and $a_{ij} = 0$.

The problem of describing the process of deformation under conditions of a complex, including cyclic loading, remains the focus of research from 1956–1957, when the first models of AYu. Ishlinsky, W. Prager, Yu.I. Kadashevich and V.V. Novozhilov were proposed, to the present. Its solution is necessary for the development of precise methods for calculating the static and long-term strength of machine parts, the development of the theory of limiting states, increasing the accuracy of calculating the shaping operations in the technological processes of forging and stamping. It is known [2] that the plastic deformation of metals in these operations is carried out under conditions of complex loading.

Over the indicated period of time, a large number of models have been proposed, which differ in their representations of isotropic and, especially, kinematic hardening under various loading schemes, including cyclic. The first group of models is based on the Ishlinsky-Prager concept of additional stress, these models differ in the form of the evolution equation for a_{ij} describing the displacement of the loading surface centre during deformation of the material [20–26] and, accordingly, the kinematic hardening.

Domestic researchers proposed various versions of the differential dependences of the monomial structure of Ishlinsky–Prager $a_{ij} = g\varepsilon_{ij}$, differing in the form of the function g [20]: $g = g(\sigma)$ – RA Arutyunyan; $g = g\left(\dfrac{(a_{ij} \cdot s^*_{ij})}{\sigma^*}, \dfrac{(\varepsilon_{ij} \cdot s^*_{ij})}{\sigma^*}\right)$ – I.A. Birger and B.F. Shorr; $g = g\left(a, \dfrac{(a_{ij} \cdot s^*_{ij})}{a \cdot \sigma^*}\right)$ – Yu.G. Korotkikh, where the sign '*' denotes the effective stresses.

Among the group of models associated with the concept of additional stress, the researchers distinguish the model of non-linear kinematic hardening by Armstrong and Frederick [21], having the form

$$\dot{a}_{ij} = \frac{2}{3} c_1 \dot{\varepsilon}_{ij} - c_2 a_{ij} \dot{\varepsilon}, \qquad (1.38)$$

where $\dot{\varepsilon} = (2/3 \dot{\varepsilon}_{ij} \dot{\varepsilon}_{ij})^{1/2}$, and c_1 and c_2 are material constants.

The two-term form of the evolution equation (1.38), in contrast to the Kadashevich–Novozhilov equation, $\dot{a}_{ij} = g(\varepsilon)\dot{\varepsilon}_{ij}$, takes into account the decrease of a_{ij} with increasing $\dot{\varepsilon}$. In addition, (1.38) is also suitable for cyclic deformation. The model (1.38) continues to improve [23].

The second group of models is the so-called multi-surface models, in which the anisotropy of the flow stress is determined by different loading surfaces [25, 26]. Modification of this direction are two-surface models based on the concept of a boundary surface in stress space [27]. The boundary surface is a stress surface located outside the yield surface. Extension of the boundary surface describes isotropic hardening, and displacement of the yield surface within the boundary describes anisotropic hardening.

A good fit to the experimental results under conditions of a complex, including cyclic, loading is shown by the model of J.L. Chaboche [23, 28]. It is the decomposition of the model (1.38) into n components, i.e.

$$\dot{a}_{ij}^{(i)} = \frac{2}{3} c_1^{(i)} \dot{\varepsilon}_{ij} - c_2^{(i)} a_{ij}^{(i)} \dot{\varepsilon}, \qquad (1.39)$$

wherein $a_{ij} = \sum_{i=1}^{n} a_{ij}^{(i)}$.

In fact, this model can be attributed to the group of multi-surface models [29].

A significant achievement, from the point of view of the description of irreversible deformation occurring under conditions of non-isothermal, non-stationary complex loading, is the theory of plastic deformation proposed by V.S. Bondar' [30]. It contains an evolution equation for an additional stress of the form

$$\dot{a}_{ij} = \frac{2}{3} g \dot{\varepsilon}_{ij} + \left(\frac{2}{3} g_\varepsilon \varepsilon_{ij} + g_a a_{ij} \right) \dot{\varepsilon}, \qquad (1.40)$$

where g, g_ε, g_a are functions determined experimentally.

The three-term structure of equation (1.40) describes the processes of formation and removal of additional stresses during plastic deformation and satisfactorily describes the hardening during deformation of specimens along various trajectories of complex loading, i.e., takes into account the vector properties of metals [20]. From a comparison of (1.40) and (1.38) it follows that the Bondar' model is a development of the Armstrong–Frederick model.

Several models are known that differ from the attempts to describe the deformation anisotropy not only by introducing additional stresses (displacement of the loading surface), but also by explicitly considering and taking into account the anisotropy parameters. The author [31, 32] proposed the loading function in the form

$$f(\sigma_{ij}) = N_{ijkl} \left(s_{ij} - a_{ij} \right) \left(s_{kl} - a_{kl} \right) - \frac{2}{3} \sigma_T^2 = 0. \qquad (1.41)$$

The fourth-rank tensor of the current anisotropy parameters N_{ijkl} is defined as

$$N_{ijkl} = J\left(\varepsilon^p\right) \left(\delta_{ik} \delta_{ji} + \delta_{jk} \delta_{il} - \frac{2}{3} \delta_{ij} \delta_{kl} \right) + \int_0^{\varepsilon^p} A\left(\varepsilon^p\right) \left(\frac{d\varepsilon_{ij}^p}{d\bar{\varepsilon}^p} \frac{d\varepsilon_{kl}^p}{d\bar{\varepsilon}^p} \right) d\varepsilon^p, \qquad (1.42)$$

where $J(\varepsilon^p)$ and $A(\varepsilon^p)$ are the scalar functions of the accumulated plastic deformation, determined experimentally.

The deviator of the additional stress is given in the form

Fundamentals of Mechanics of Strength and Plasticity

$$a_{ij} = \frac{2}{3}\sigma_T \int_0^{\varepsilon^P} B(\varepsilon^P)\frac{d\varepsilon_{ij}^p}{d\overline{\varepsilon}^p}d\varepsilon^P, \tag{1.43}$$

where $B(\varepsilon^p)$ is the experimentally determined scalar function.

In this case, $\varepsilon^p = \int d\overline{\varepsilon}^p$, $d\varepsilon_{ii} = 0$ and the associated flow law takes the form

$$d\varepsilon_{ij}^p = d\lambda \frac{\partial f}{\partial \sigma_{ij}} = d\lambda N_{ijkl}(s_{kl} - a_{kl}). \tag{1.44}$$

From the resulting equation it follows that in the process of deformation of the body element, the loading surface expands, shifts, rotates and changes its shape. In this case, its final form and position depend on the loading history. This corresponds to the available experimental data, which were obtained by the author himself.

Figure 1.6 shows the projections on the plane $\sigma_x + \sigma_y = 0$ and $\tau_{xy} = 0$ of the section of the plasticity ellipsoid (plane stress state) in the stress space σ_x, σ_y and τ_{xy}

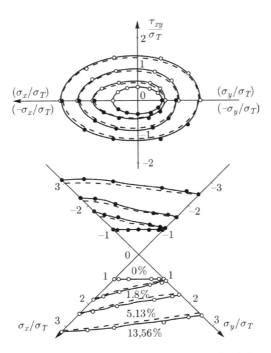

Fig. 1.6. Changing the ellipsoid cross section of the plasticity of copper M1 with increasing degree of deformation: points – experiment; solid lines – calculation based by the author's model [31, 32]; dashed – calculation by the Kadashevich–Novozhilov model.

$$\sigma_x^2 + \sigma_y^2 - \sigma_x\sigma_y + 3\tau_{xy}^2 = \sigma_T^2 \qquad (1.45)$$

plane $\sigma_x + \sigma_y = \sigma_T$. The ellipsoids describe the state of a copper M1 sample (isotropic in the initial state), deformed along the trajectories of complex loading by the total degree of deformation ε^p = 1.8; 5.13; 13.56% for subsequent stretching (light points) and draft (black points). It is seen that in the process of deformation the plasticity curve widens, noticeably shifts to the left and less noticeably upward, i.e., an isotropic and complex translational hardening occurs. In this case, the deformation law (1.44) describes the experimental data more accurately, and the law (1.35) – with an error not exceeding 8%. With an increase in the degree of deformation the error does not increase. However, this insignificant increase in accuracy is achieved by a significant complication of both the equations and the basic experiment for determining the material functions.

1.2.2. Scalar defining relations

Using the tensor laws of metal deformation (1.21), (1.26), (1.29), etc., to calculate the stress–strain state of a billet processed by pressure metal working methods, it is necessary to have defining scalar relations describing the dependence of the deformation resistance of a particular material on the degree, strain rate and temperature:

$$\sigma = f(\varepsilon, \dot\varepsilon, T). \qquad (1.46)$$

The empirical approach was most widely used to determine dependences (1.46) [36–39]. Tests of cylindrical standard samples are carried out by stretching, torsion or compression (upset) on universal testing machines or cam plastometers with varying temperature and strain rate. The true diagram of deformation is most simply constructed from the results of testing specimens for compression. To exclude contact friction forces on the ends of the samples, cylindrical or conical depressions are carried out, which are filled with grease before upsetting by flat ground strikers. The procedure for carrying out tests and calculations is described in detail, for example, in [2, 38].

With cold deformation at moderate rates at which dynamic effects can be neglected, the flow stress is not sensitive to the rate of deformation and the experimental results are approximated by power equations:

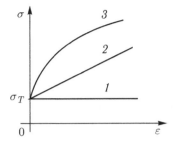

Fig. 1.7. Schematicized deformation diagrams: 1, 2 and 3 – rigid perfectly plastic body and bodies with linear and non-linear hardening.

$$\sigma = \sigma_T + b\varepsilon^n; \qquad (1.47)$$

$$\sigma = c\varepsilon^n, \qquad (1.48)$$

where b, c and n are the phenomenological coefficients determined from the experimental deformation diagrams.

When hot deformation is used as a rule, the method of experiment planning and the dependences (1.46) are, for example, in the form [39]

$$\sigma = a\dot{\varepsilon}^m \cdot \varepsilon^n \exp(-bT). \qquad (1.49)$$

When solving practical problems of determining the stress–strain state (stress and strain fields) in the case of plastic deformation of a body, schematization of the experimental dependences $\sigma(\varepsilon)$ of specific materials is used as a rule. The most frequently used schematics shown in Fig. 1.7. In this case, when calculating large plastic strains the strain elastic components are usually neglected because of their smallness in comparison with the residual strains, i.e., material deformation diagrams are used.

The above equations do not exhaust the rich variety of formulas proposed in the literature for approximating the experimental dependences of metal deformation resistance on thermomechanical parameters. A rather detailed review of them is given in [38]. Being purely phenomenological, they do not disclose the physical content of the influence of ε, $\dot{\varepsilon}$ and T on the resistance to deformation and, therefore, do not allow purposefully to design the loading function and to obtain the tensor (generalized) deformation law adequate to the deformation behaviour of specific materials.

In addition, the above equations do not take into account the effect on σ of loading history associated with the ε, $\dot{\varepsilon}$ and T dependences on time. Plastic deformation is a non-equilibrium and irreversible process of transition of a metal as a system from one state to another.

Therefore, its final state is determined not only by the initial state, but also depends on the path along which this transition was carried out. Hence it follows that the deformation resistance in the general case can not be described by the functional dependence (1.46), but is a functional of the form [2]

$$\sigma = F\left[\varepsilon(t), \dot{\varepsilon}(t), T(t), \rho_i(t), t\right], \qquad (1.50)$$

where $\rho_i(t)$ are the characteristics of the material structure at different levels.

Elementary analysis (1.50) shows, firstly, that it is very difficult to develop and build a test machine for the experimental determination of σ, taking into account all factors affecting it, since it is necessary to control a large number of functions during the deformation of samples, among which the most difficult to control functions of the structure parameters $\rho_i(t)$ are represented. Secondly, even if the functions (1.50) are neglected by the arguments $\rho_i(t)$, then the construction of a system of control and monitoring of the parameters of the test machine, which is very complicated by the kinematic scheme, would not be meaningful either. The point is that, for example, industrial pressure metal working technological processes are characterized by the presence of inhomogeneous and nonstationary tensor fields of plastic strain rates $\dot{\varepsilon}_{ij}(x, y, z, t)$, deformation $\varepsilon_{ij}(x, y, z, t)$ in the volume of the workpiece being machined, and the scalar temperature field $T(x, y, z, t)$. Therefore, practically infinite sets of dependences $\varepsilon(t)$, $\dot{\varepsilon}(t)$ and $T(t)$ and their combinations can not be realized in the experiment, in accordance with (1.50). There remains only one way – the theoretical construction of the functional (1.50).

The authors of [40] made an attempt to construct a functional within the framework of the phenomenological approach based on the creep theory of Yu.N. Rabotnov [41]. The result is as follows:

$$\sigma(t) = \varphi\left[\varepsilon(t)\right] - \int_0^t R(t-\tau)\varphi\left[\varepsilon(\tau)\right] d\tau, \qquad (1.51)$$

where $0 < \tau < t$; $\varphi[\varepsilon(t)]$ and $R(t - \tau)$ are experimentally determined functions.

The first term in (1.51) describes the deformation resistance due to hardening only. To determine it, the deformation of the samples must be carried out with large $\dot{\varepsilon}$, when it can be assumed that softening processes do not have time to proceed. The second term describes the processes of softening and can be determined from softening tests.

Fundamentals of Mechanics of Strength and Plasticity

The functional (1.51) does not take into account the effect on the strain resistance of the dependences $T(t)$, $\dot{\varepsilon}(t)$ and $\rho_i(t)$. The experiment on finding the above functions is also difficult and time-consuming.

In conclusion, one should quote an excerpt from [2]: "The construction of functionals (more precisely, operators) that describe the resistance of metals to plastic deformation in all the complexities of this phenomenon is a major scientific problem."

1.3. Fundamentals of the classical mathematical theory of creep of metals

In addition to plastic (active) deformation, another type of irreversible deformation of materials loaded with external forces is observed, which is of great importance for engineering and is called *creep deformation*. The physical phenomenon of material acquisition at an elevated temperature, increasing with time, irreversible deformation at stresses less than the yield point is called *creep* or *aftereffect*. The other side of this phenomenon is *stress relaxation* – a gradual decrease with time of stresses in an elastically deformed and fixed sample.

Creep has become the object of intensive research since the late XIX–early XX century in connection with the development of industrial production. The construction of boilers and other chemical and metallurgical equipment operating at elevated temperatures and subject to moderate loads required information on the behaviour of

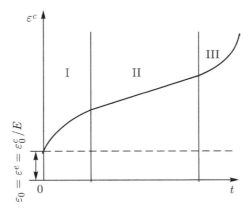

Fig. 1.8. A typical creep curve: I – the first unsteady stage; II – the second steady-state stage, at which $\dot{\varepsilon} = d\varepsilon^c / dt = \text{const}$; III – the third stage.

28 Physico-Mathematical Theory of High Irreversible Strains

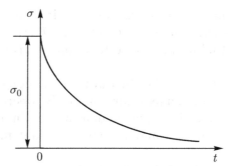

Fig. 1.9. A typical stress relaxation curve.

metals under these conditions. This required knowledge of both the stress–strain state of the parts, and the evolution of their structure, and consequently, of the mechanical properties.

The results of the studies in the first five to six decades were summarized in monographs [42, 43].

Here we do not set out the goal of giving a detailed review of the development of creep theory and its current state. Our task is to formulate the basic concepts, the regularities of the phenomenon, the realized approaches to its theoretical description and the main results obtained, as well as the emerging new approaches to the further development of the theory and the problems that have not been solved until the end.

The basic law of creep is described by the *creep curve* – the dependence of the creep strain ε^c on the time of the sample under load (usually at a constant stress $\sigma^c < \sigma_T$ at a constant temperature) (Fig. 1.8). Creep curves are constructed from the test results (usually tensile loading) of the samples. The sample is placed in the furnace of a special testing machine having the required temperature, loaded under the condition $\sigma^c = $ const $< \sigma_T$ or tensile force $P = $ const. At certain fixed intervals, the deformation acquired by the sample is measured.

The law of stress relaxation is the *relaxation curve* – the dependence of stress σ in a preloaded sample versus time under the condition of constant deformation, which it acquired under the action of the initial stress σ_0 at $t = 0$ (Fig. 1.9).

For experimental detection of creep patterns, the tests of the samples are carried out at different T and σ, including at $T(t)$ and $\sigma(t)$.

The complete theory of creep, like the theory of plasticity, must contain the phenomenological and physical parts. The latter is necessary to understand the laws of the phenomenon and, possibly, the prediction of new effects.

To solve applied (engineering) problems, the theory should give generalized creep laws, that is, the equations of the connection between tensor stress and strain fields in detail, working including under conditions of non-stationary creep, and also predicting the limiting state – metal destruction and the accumulated creep strain.

We will try to consider the existing theory of creep in the light of this set of requirements.

Already in the first experimental studies, the main regularities of creep deformation were revealed, related to the creep rate $\dot{\varepsilon}^c = d\varepsilon^c/dt$ as a function of temperature and stress. It became clear that the phenomenon is due to the viscosity of metals, which is quite noticeable at elevated T and, consequently, is controlled by diffusion in metals. There was a natural division of research into two groups: using phenomenological and structural (physical) approaches [44]. Both approaches were based on the results of an empirical, i.e., experimental study of the phenomenon.

The first quantitative description of creep was proposed in the form of a power series [44, 45]

$$\dot{\varepsilon}^{\tilde{n}} = \sum_i a_i t^{-n_i}, \qquad (1.52)$$

where a_i and n_i are functions of T and σ^c.

The value of n_i for most metals lies in the range $0 \le n_i \le 1.0$, and with increasing T the coefficient n_i decreases. At low *homological temperatures*[1], when the recovery does not practically develop, and at large σ^c, a logarithmic law of uniaxial creep is observed:

$$\varepsilon^c = a\ln(\gamma t + 1) + \varepsilon_0. \qquad (1.53$$

Equation (1.53) is obtained by integrating the monomial equation (1.52) for $n = 1.0$. In it, a and γ are constants. It is interesting to note that equation (1.53) is simultaneously a solution of the differential equation

$$\dot{\varepsilon}^c = a\gamma \exp\left(-\frac{\varepsilon^c - \varepsilon_0}{a}\right).$$

The equations of this type for $a \simeq kT$, where k is Boltzmann's constant, describe thermally activated processes with an activation

---
[1] Homological temperature (similar) $T_\theta = T/T_m$, where T is the deformation observation temperature, T_m is the melting point of the material.

energy proportional to ε^c.

At high T_θ, when thermal activation and recovery play a predominant role, the creep curve $\varepsilon_c(t)$ is described by the power equation derived by E.N. Andrade [44]:

$$\varepsilon^c = \beta t^{1/3} + \varepsilon_0, \qquad (1.54)$$

where β is a constant coefficient.

At the steady-state second stage (Fig. 1.8), when the creep rate is minimal and constant,

$$\dot\varepsilon^c_{\min} = d\varepsilon^c/dt = const,$$

the deformation is described by a single-term equation (1.52) for $n = 0$,

$$\varepsilon^c = \dot\varepsilon^c_{\min} \cdot t + \varepsilon_0. \qquad (1.55)$$

The steady-state creep, which occurs at σ, T = const, is proposed to be considered as a result of the dynamic equilibrium between strain hardening and thermally activated recovery.

Equation (1.55) for $\varepsilon_0 = 0$, very small σ^c and high temperatures, describes the Nabarro–Herring and Coble diffusion creep [45]:

$$\varepsilon^{\tilde{n}} = \dot\varepsilon \cdot t = \sigma / \mu \cdot t, \qquad (1.56)$$

where $\dot\varepsilon = \sigma^{\tilde{n}}/\mu$; μ is the coefficient of viscosity of the material.

At these strains, mass transfer is carried out atomically (individual atoms). A more general form of creep is dislocation creep, in which mass transfer and, consequently, irreversible deformation are effected by the motion of dislocations. The dislocation creep will be the object of our consideration. It is most often observed in the operation of machines and structures.

Creep curves are the basis for calculating machine parts for creep. As characteristics of the resistance of the creep material, two versions of the concept of creep are introduced [7]: 1) *the creep limit $\sigma^c_{\lim}$ is the conditional stress at which the creep strain within a specified time interval reaches the value established by the technical conditions;* 2) *the creep limit is the stress at which the creep strain rate is equal to a certain value established by the technical conditions.*

To solve practical engineering problems related to the strength calculations of machine parts and constructions on creep, three-

Fundamentals of Mechanics of Strength and Plasticity 31

dimensional laws are necessary, in addition to the one-dimensional creep laws considered. The creep strain, like the plastic deformation, is irreversible. Therefore, the construction of the phenomenological creep theory is carried out using a method analogous to the method of constructing the theory of plasticity, ie, the applicability of the basic hypotheses of the theory of plasticity is postulated.

In particular, the existence of a finite set of determining parameters and corresponding potentials, for example, the creep strain rate potential f^c, is assumed. Then

$$\dot{\varepsilon}_{ij}^c = \lambda \frac{\partial f^c}{\partial \sigma_{ij}}. \tag{1.57}$$

Substituting (1.57) into the expression for the intensity of the strain rates

$$\dot{\varepsilon}^n = \sqrt{2/3\, \dot{\varepsilon}_{ij}^c \cdot \dot{\varepsilon}_{ij}^c}$$

the value of λ is determined:

$$\lambda = \sqrt{\frac{3}{2}} \frac{\dot{\varepsilon}^c}{\sqrt{\frac{\partial f}{\partial \sigma_{ij}} \bigg/ \frac{\partial f}{\partial \sigma_{ij}}}}. \tag{1.58}$$

The equation $f^c = 0$ is the equation of the hypersurface in the space of components of the stress tensor σ_{ij}, to which, according to (1.57), the creep strain velocity vectors at the corresponding points σ_{ij} are orthogonal. It is called the *creep hypersurface*. If the isotropy of the initial material is accepted, isotropic hardening, the volume invariance $\varepsilon_{ii}^c = 0$ or $\dot{\varepsilon}_{ii}^c = 0$, and the independence of f^c only from the second invariant of the stress tensor, then

$$\frac{\partial f}{\partial \sigma_{ij}} = \frac{\partial f}{\partial s_{ij}} = 3 s_{ij}$$

and according to (1.58)

$$\lambda = \frac{1}{2} \frac{\dot{\varepsilon}^c}{\sigma^c}.$$

Consequently, the defining creep relations take the form

$$\dot{\varepsilon}_{ij}^c = \frac{3}{2} \frac{\dot{\varepsilon}^c}{\sigma^c} \left(\sigma_{ij} - \delta_{ij} \sigma_0 \right). \tag{1.59}$$

32 Physico-Mathematical Theory of High Irreversible Strains

We outlined a general method for constructing the theory of creep. It follows from this that various versions of the theory are determined by various options for choosing the creep potential.

The so-called *technical creep theories* were widely used in creep calculations. They differ in the set of determining parameters used to describe creep and the type of dependence between them. These are the aging theories $\sigma^c = \varepsilon^c = \Phi_1(\varepsilon^c, t)$, the flow $\sigma'' = \Phi_2(\dot{\varepsilon}'', t)$, and the hardening $\sigma^c = \Phi_3(\dot{\varepsilon}^c, \overline{\int d\varepsilon}^c)$ [7]. In the ageing theory, the creep potential depends on the second invariant of the stress deviator, strain intensity ε^c and time t:

$$f_1^c = \frac{3}{2} s_{ij} s_{ij} - \left[\Phi_1(\varepsilon^c, t)\right] = 0.$$

Then

$$\varepsilon_{ij}^c = \lambda \frac{\partial f_1^c}{\partial \sigma_{ij}} = \frac{3}{2} \frac{\varepsilon^c}{\sigma^c} \left(\sigma_{ij} - \delta_{ij} \sigma_0\right). \qquad (1.60)$$

It follows from (1.60) that the solution of the problem of determining the stress–strain state of a part for a certain time value with the application of the theory of ageing is equivalent to solving the problem of the theory of small elastoplastic strains with known deformation diagrams – *isochronous creep curves* for different values of t.

The theory of ageing, providing for the existence of a definite relationship between σ^c, ε^c and t, was proposed by C.R. Soderberg and generalized by Yu.N. Rabotnov [7].

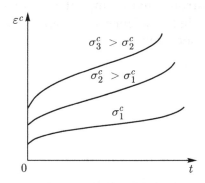
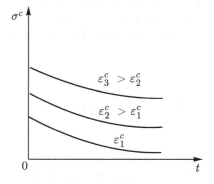

Fig. 1.10. Cross sections of the creep surface in the coordinates ε^c, σ^c, t by planes perpendicular to the σ^c axis – creep curves.

Fig. 1.11. Cross sections of the creep surface by planes perpendicular to the axis ε^c – relaxation curves.

Fundamentals of Mechanics of Strength and Plasticity

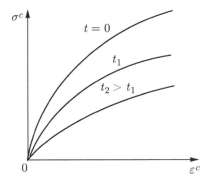

Fig. 1.12. The sections of the creep surface by planes perpendicular to the t axis – isochronous creep curves.

The existence of the above dependence is equivalent to the assumption of existence for a given T of the surface in the coordinates σ^c, ε^c, and t. Successively dissecting this surface by systems of parallel planes perpendicular to the axes σ^c, ε^c and t, we obtain respectively the creep curves $\varepsilon^c(t)$ and $\sigma^c = $ const, the stress relaxation curves $\sigma^c(t)$ and $\varepsilon^c = $ const, as well as the $\sigma^c(\varepsilon^c)$ of the values of t (Figs. 1.10, 1.11 and 1.12, respectively) [7].

The dependences $\sigma^c(\varepsilon^c)$ in Fig. 1.12 are called isochronous creep curves. To calculate the component for creep, i.e., to determine the tensor fields (by volume of the component) of stresses and strains for a particular time of operation of the compo,ent using the theory of ageing, it is necessary to formulate mathematically a problem that will include differential equilibrium equations, Cauchy's kinematic relations and defining equations (1.60). When solving a system of equations of a problem, for example, using computer simulation software, it is necessary to use an isochronous creep curve for a given time as a deformation diagram in solving plastic problems.

If it is necessary to take into account the elastic component of the deformation, the defining relations must be used in the form

$$\varepsilon_{ij} = \frac{1}{2G}(\sigma_{ij} - \delta_{ij}\sigma_0) + \frac{3}{2}\frac{\varepsilon^c}{\sigma^c}(\sigma_{ij} - \delta_{ij}\sigma_0) = \frac{3}{2}\frac{\varepsilon}{\sigma}(\sigma_{ij} - \delta_{ij}\sigma_0), \quad (1.61)$$

where G is the shear modulus; $\varepsilon = \varepsilon^e + \varepsilon^c$ is the total strain rate including elastic ε^e and creep strain ε^c.

According to the flow theory, there is a relationship between σ^c, $\dot\varepsilon^c$ and t. The creep potential is taken in the form

$$f^c = \frac{3}{2}s_{ij}s_{ij} - \left[\Phi_2(\dot{\varepsilon}^c, t)\right]^2 = 0,$$

and the defining equations will have the form

$$\dot{\varepsilon}_{ij}^c = \frac{3}{2}\frac{\dot{\varepsilon}^c}{\sigma^c}(\sigma_{ij} - \delta_{ij}\sigma_0). \tag{1.62}$$

For the application of (1.62) in the calculation of workpieces for creep, it is necessary for a particular material to have a set of dependencies $\sigma^c(\dot{\varepsilon}^c)$ for different values of time t.

The flow theory was proposed by C.C. Davenport and received practical application thanks to the work of L.M. Kachanov [7].

In the hardening theory in the case of a non-uniform stress state, the existence of a relationship between stress intensities, creep strain during time t, and creep strain rate is assumed, as before. The equation of the potential surface is taken in the form

$$f^c = \frac{3}{2}s_{ij}s_{ij} - \left[\Phi_3\left(\dot{\varepsilon}^c, \int q^c\right)\right]^2 = 0.$$

In this case the defining relations have a form analogous to (1.62). The difference lies in the fact that in solving the problems of hardening theory, it is necessary to have $\sigma^c(\varepsilon^c)$ for various $\dot{\varepsilon}^c$ to use equations (1.62). Often this dependence is approximated by the analytical equation

$$\sigma = \left[\frac{\dot{\varepsilon}^c}{\alpha}(\varepsilon^c)^\beta\right]^{1/\nu}, \tag{1.63}$$

where α, β and ν are the coefficients depending on T, determined for the material under consideration from the experimentally obtained dependences $\sigma^c(\varepsilon^c)$ for various $\dot{\varepsilon}^c$.

The hardening theory was proposed by P. Ludwik, A. Nadai and W.H. Davenport and developed by Yu.N. Rabotnov and S.A. Shesterikov [7].

Technical creep theories have been used for many years in the calculation of machine parts and structures. However, with a sufficient accuracy for practice, only problems in a simple formulation are solved. As a rule, the process of steady-state creep (stresses are constant) is considered at a constant temperature.

It is known that the scope of any theory is determined by the hypotheses and simplifying assumptions adopted in its construction. The main drawback of the technical theories of creep is the impossibility of a correct description of non-stationary (unsteady processes), when all the thermomechanical loading parameters that are related together change over time, that is, $\sigma^c(t)$, $\varepsilon^c(t)$, $\dot{\varepsilon}^c(t)$, $T(t)$. This case is not described by any of the above creep theories. A number of approximate methods have been developed for the solution of fairly simple (in the sense of geometry and the system of external forces) problems of unsteady creep with the application of technical creep theories in an isothermal formulation [7]. A method of successive approximations (A.A. Il'yushin, I.I. Pospelov), the principle of a minimum of additional power (L.M. Kachanov) are also proposed.

Another major problem of the phenomenological theories of creep, as well as the mathematical theory of plasticity, is the problem of taking into account the influence on the process of metal structure and its evolution during the work of the parts.

In creep metals exhibit the whole set of properties, due to which they are the main structural material in engineering; this is the elasticity, ductility and viscosity at high strength. Therefore, the problem of creep can be designated as a problem of a general elastic-viscoplastic body [46].

Since the end of the first half of the last century, the efforts of researchers have been aimed at improving the theory and, on its basis, the methods of design calculations. A great contribution to the development of creep theory was made by the Russian scientists A.A. Il'yushin, I.I. Pospelov [47], L.M. Kachanov [48], N.N. Malinin [7, 49], Yu.N. Rabotnov [41], O.V. Sosnin [50], S.A. Shesterikov [51], A.M. Lokoshchenko, V.P. Radchenko [52, 53] and others.

The processes of irreversible deformation are non-equilibrium, so the final state of the system is determined not only by the finite values of the thermodynamic parameters, but also by the law of their variation upon transition of the system from the initial state to the final state [54].

In the case of irreversible deformation of metals, these laws are called the load history. The development of the theory of irreversible deformation from the end of the first half of the twentieth century was determined by the desire to take into account the history of loading and deformation.

To this end, Yu.N. Rabotnov proposed including in the creep potential of the theory of ageing, in addition to the intensity of deformation and time, other internal variables, which were called *structural parameters*, and also introduce kinetic equations describing the evolution of these parameters [41]. The idea turned out to be fruitful.

Yu.N. Rabotnov and L.M. Kachanov proposed the use of the notion of material damage introduced by them and the kinetic equation for its description as a structural parameter. This equation, when added to the creep equations, made it possible to describe the long-term strength of materials [55, 56]. These works served as the basis for the formation of a new direction in mechanics, the mechanics of continual damages, or the kinetic theory of creep and long-term strength [53]. Another approach to the problem of long-term strength is the criterial approach [52]. We will touch on it below.

Macrocharacteristics have been proposed to describe the creep of metals directly as structural parameters: the Udquist parameter, the work of plastic deformation, etc. Since the structural parameters must remain unchanged under the formula (1.57), the calculation of the loading history turns out to be problematic.

Another attempt to solve the problem of accounting for the history of loading was the *theory of heredity*. The simplest version of the linear theory, based on the principle of superposition of strains, was proposed by Boltzmann [7]. Let us briefly consider the logic of this approach, following [7]. Let the stress in a stretched rod be constant and equal to $\sigma(\xi)$ at the instant of time ξ for a small time interval $d\xi$. This stress will cause a time-varying deformation. Let us assume that at a time $t > \xi$ this deformation is proportional to the stress $\sigma(\xi)$, the time of its action $d\xi$ and some decreasing function of the time interval $t-\xi$, which we denote by $H(t - \xi)$, and inversely proportional to the elasticity modulus E. The decrease of the function H is due to the 'forgetting' of the material with the passage of time t of the action σ. The argument of a function H of the form $(t-\xi)$ means its invariance with respect to the origin of time.

According to the principle of superposition, the value of the deformation at time t, which arose from stresses that operated before time t, is $\dfrac{1}{E}\int\limits_0^t H(t-\xi)\sigma(\xi)d\xi$.

The stress at time t initiates an elastic strain σ/E. The total strain is made up of elastic strain and strain which arose from the stress acting up to the time t:

$$\varepsilon^c = \frac{1}{E}\left[\sigma + \int_0^t H(t-\xi)\sigma(\xi)d\xi\right]. \tag{1.64}$$

This equation allows, according to the well-known law $\sigma(\xi)$, to describe the creep phenomenon at a constant initial stress $\sigma(0) = $ const. In this case we have

$$\varepsilon^c = \frac{\sigma(0)}{E}\left[1 + \int_0^t H(t-\xi)d\xi\right]. \tag{1.65}$$

To use (1.65) in the calculations, it is necessary to know the kernel $H(t-\xi)$ of the integral equation. In this case (1.65) describes the creep curve for $\sigma_0 = $ const. The core is determined from the experimental creep curves for the material in question.

In the case of a non-uniform stress state in (1.64), it is necessary to pass from the intensities to the tensors:

$$e_{ij} = \frac{1}{2G}\left[s_{ij} + \int_0^t H(t-\xi)\sigma(\xi)d\xi\right], \tag{1.66}$$

where e_{ij} is the deviator of the strain tensor.

In this case, the solution of the creep problem reduces to solving an elastic problem with varying boundary conditions, in which the elastic constants are replaced by rheological operators. This result is called the Volterra principle. Problems must be formulated in an

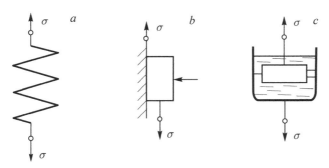

Fig. 1.13. Mechanical (rheological) models of the simplest ideal bodies: elastic (*a*), plastic (*b*) and viscous (*c*).

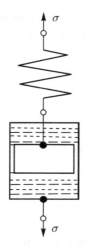 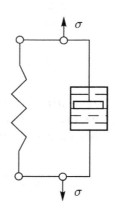

Fig. 1.14. Maxwell's body model – a serial connection of elastic and viscous elements.

Fig. 1.15. Voigt body model – a parallel connection of elastic and viscous elements.

isothermal setting, that is, the problem of an elastoviscoplastic body in the general case remains unresolved.

A sufficiently developed direction of the theory of plasticity and creep is the use of rheological (mechanical) models to describe the basic properties of metals. The method is based on three elementary elements, modelling the elasticity, ductility and viscosity of materials [7, 8]. The first element is an elastic spring (Fig. 1.13a), the deformation of which is described by the uniaxial Hooke's law

$$\sigma = E\varepsilon_{el}. \qquad (1.67)$$

The element describing the ideal plasticity, $\sigma = \sigma_T$, is a solid body sliding along a rough surface (Fig. 1.13, b).

The element modelling the viscosity consists of a cylinder into which a piston is inserted with a clearance, and the cylinder is filled with a viscous liquid. The uniaxial law of deformation of an element is the equation of Newton's ideal viscosity fluid

$$\sigma = \mu\dot\varepsilon, \qquad (1.68)$$

where μ is the viscosity coefficient.

The models are loaded with tensile stresses. To simulate a set of properties (simple and complex bodies), the simplest models are connected in series and in parallel (Fig. 1.14).

Fundamentals of Mechanics of Strength and Plasticity

It is assumed that in a sequential connection, the deformation of the composite model is equal to the sum of the deformations of each element, and the stress across the elements is the same. When the elements are connected in parallel (Fig. 1.15), the stress of the composite model is equal to the sum of the stresses necessary to deform each element, and the deformation of all elements is the same.

These rules make it easy to obtain rheological equations for the deformation of complex bodies. For example, for Maxwell's body (Fig. 1.14), we have

$$\varepsilon_M = \varepsilon_{el} + \varepsilon_{vis}. \tag{1.69}$$

Differentiating (1.69) with respect to time and using the rheological equations of elastic and viscous bodies, we obtain the law of deformation of Maxwell's viscoelastic body:

$$\dot{\varepsilon}_M = \dot{\varepsilon}_{el} + \dot{\varepsilon}_{vis} = \frac{1}{E}\frac{d\sigma}{dt} + \frac{\sigma}{\mu}. \tag{1.70}$$

For a constant strain $\dot{\varepsilon}_i = 0$ (the sample was stretched in elastic limits and fixed), it follows from (1.70) that

$$\frac{1}{E}\frac{d\sigma}{dt} + \frac{\sigma}{\mu} = 0.$$

Integrating this equation with the initial condition $\sigma = \sigma_0$ for $t = 0$, we obtain

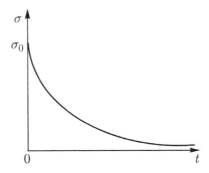

Fig. 1.16. The stress relaxation curve in Maxwell's body.

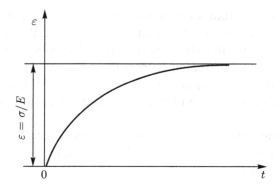

Fig. 1.17. The creep curve of Voigt's body at σ = const.

$$\sigma = \sigma_0 \exp\left(-\frac{t}{t_0}\right), \qquad (1.71)$$

where $t_0 = \mu/E$ is the time over which the stress σ_0 decreases by $e = 2.718$ times. This quantity is called the *relaxation time*.

The graphical form of the dependence $\sigma(t)$ (1.71) is shown in Fig. 1.15. It follows from this that the model of Maxwell's body, with proper selection of the quantities E and μ, can describe the stress relaxation curves of real metals.

Now we connect the elastic and viscous elements in parallel (Fig. 1.16). The model stress is

$$\sigma = \sigma_{el} + \sigma_{vis} = E\varepsilon_{el} + \mu\dot{\varepsilon} = E\varepsilon_{el} + \mu\frac{d\varepsilon}{dt}, \qquad (1.72)$$

Integrating (1.72) with σ = const and the initial condition $\varepsilon = 0$ for $t = I = 0$, we obtain

$$\varepsilon = \frac{\sigma}{E}\left[1 - \exp\left(-\frac{E}{\mu}t\right)\right]. \qquad (1.73)$$

A graphical depiction of the dependence (1.73) is given in Fig. 1.17. It is clear that with an appropriate selection of the values of E and μ, equation (1.73) can describe the creep curve of a real metal.

We outlined a general approach to the construction of creep models and plasticity using mechanical models of the basic mechanisms of metal deformation. Generalization of uniaxial rheological models to a volumetric stress state is accomplished by changing the intensities with the corresponding tensors. More detailed information on the method can be obtained in [8]. Here, in accordance with the main purpose of the above review, we note the following. Despite the external and internal beauty of the method, it does not solve the problem of an elastoviscoplastic body. Rheological coefficients, for example, μ in (1.7) and in (1.73), are unknown functions of temperature, stress, the characteristics of the metal structure, and also the history of loading and deformation. Experimental determination is possible for a limited number of simple deformation schemes.

Other approaches to the development of creep theory will be considered below. We note that, similarly to the theory of plasticity, the use of the concept of additional stress (see section 1.2) makes it possible to take into account the strain anisotropy in the creep theory [7].

1.4. Modern approaches to the development of the mathematical theory of irreversible strains and the formulation of a scientific problem

1.4.1. Plasticity theory

Over the years, the main trend in the development of the plasticity theory, following from practical inquiries, was the generalization of the flow theory to large (finite) elastoplastic and elastoviscoplastic strains [57–59]. The movement in this direction within the framework of a purely phenomenological approach is connected with the solution of three main tasks. The first is to develop a procedure for dividing the strains into reversible (elastic) and irreversible (plastic), due to structural changes in the material. This is a purely kinematic problem.

The second problem is related to the definition of the tensor of the rate of change of irreversible strains, with the help of which the associated plastic flow law is formulated (1.17). Separation of strain by any of the numerous criteria into reversible and irreversible components determines the problem of choosing an objective derivative for determining the rates of plastic strains. This problem is also associated with the development of kinematics of large elastoplastic strains.

The third problem is due to the correct consideration of the viscosity of materials in the law of irreversible deformation and, ultimately, taking into account the loading history.

Some successes in this direction are related to the work of E.H. Lee [60], VI. Levitas [57], L.V. Kovtanyuk [58], G.I. Bykovtsev and A.V. Shitikov [61], V.P. Myasnikov, A.A. Burenin [62, 63], B.V. Annin and V.M. Zhigalkin [64]. and others.

A rather detailed review of the current state of the problem of constructing the theory of large elastoplastic strains is given in one of the recent papers devoted to this problem [58]. Consider briefly, following [58], the state of the problem and the solutions proposed.

One of the first hypotheses in the construction of the flow theory is the assumption of a complete deformation of an element of a continuous medium in the form of a sum of reversible (elastic) and irreversible (plastic) components: $\varepsilon_{ij} = \varepsilon_{ij}^e + \varepsilon_{ij}^p$. From the mathematical and physical points of view, this equality is fairly accurately fulfilled in the range of small strains ($\varepsilon \leq 0.1 \div 0{,}2$). On the other hand, only total strains can be measured experimentally. This implies the problem of determining ε_{ij}^e and ε_{ij}^p for large strains, which causes the problem of 'choosing' the objective derivative of irreversible deformation in time.

Without dwelling on the solutions of these problems proposed by different authors (they were discussed in [58]), we briefly consider the author's proposal [58].

The algorithm for solving the problem of determining reversible e_{ij} and irreversible p_{ij} strains is as follows.

1. The Almansi strain tensor d_{ij} is taken as a measure of deformation, and, as in the case of small strains, the following sum is considered correct:

$$d_{ij} = e_{ij} + p_{ij}. \tag{1.74}$$

2. The Lee condition [60] is clarified and is based on the hypothesis that to each deformed state with accumulated p_{ij} there corresponds a single unloading state, independent of the nature of the unloading process. It is toughened by the condition that during the unloading process all changes in the components of the tensor p_{ij} are related only to a rigid displacement and rotation of the coordinate system. In this case, as shown by the appropriate calculations, the tensors e_{ij} and p_{ij} become independent thermodynamic parameters of the process and are determined by the differential equations of their change (transport), having the form

$$d\dot{p}_{ij} = r_{ik} p_{kj} + p_{ik} + r_{jk}, r_{ij} = -r_{ji}, \tag{1.75}$$

Fundamentals of Mechanics of Strength and Plasticity

$$d\dot{e}_{ij} = \dot{\varepsilon}_{ij} + \frac{1}{2}\left(r_{ik}e_{kj} - e_{ik}r_{kj} - v_{k,i}e_{kj} - e_{ik}v_{k,j}\right), \quad (1.76)$$

where v_k are the components of the velocity vector; $\dot{\varepsilon}_{ij}$ is the strain rate tensor; r_{ik} is the skew-symmetric tensor.

3. On the basis of the well-known connection between the Almansi tensor d_{ij} and the metric tensor $g_{ij} = a_{k,i}\, a_{k,j}$, where a_k are the initial (material) coordinates of the form

$$d_{ij} = \frac{1}{2}\left(\delta_{ij} - g_{ij}\right), \quad (1.77)$$

δ_{ij} are the components of the unit tensor, for the required expansion the following dependence is obtained

$$d_{ij} = e_{ij} + p_{ij} - \frac{1}{2}e_{ik}e_{kj} - e_{ik}p_{kj} - p_{ik}e_{kj} + e_{ik}p_{km}e_{mj}. \quad (1.78)$$

4. Further, based on the law of conservation of energy, the dynamics of the deformation process is considered. The defining relations for the reversible deformation region ($p_{ij} = 0$) are obtained in the following form:

$$\sigma_{ij} = \rho \frac{\partial \psi}{\partial d_{ik}}\left(\delta_{kj} - 2d_{kj}\right), \quad (1.79)$$

where ρ is the density of the medium; $\psi = e - T$ is the thermodynamic potential (density of distribution of free energy); e is the mass density of the internal energy distribution; σ_{ij} are the components of the Cauchy tensor; T is the thermodynamic temperature.

Since the stresses are uniquely determined by elastic strains in the unloading processes and under the plastic flow, the elastic strains are related to the stresses by the dependence

$$\sigma_{ij} = \rho \frac{\partial \psi}{\partial e_{ik}}\left(\delta_{kj} - e_{kj}\right). \quad (1.80)$$

A direct consequence of the thermodynamic approach is the definition of an objective time derivative that relates irreversible strains to the rate of their change:

$$\dot{\varepsilon}_{ij}^p = \frac{Dp_{ij}}{Dt} = \frac{dp_{ij}}{dt} + p_{is}r_{sj} + p_{is}\dot{\varepsilon}_{sj}^p + r_{si}p_{sj} + \dot{\varepsilon}_{si}^p p_{sj}. \quad (1.81)$$

Consequently, (1.81) is not the subject of a 'choice' of the existing objective derivatives (Jauman, Cotter–Rivlin, Trusdell, etc.), as in most of the proposed generalizations of the flow theory, but follows directly from the laws of thermodynamics, that is, the solution of the second problem logically follows from the solution of the first.

The generalized plastic flow theory is constructed using the basic method of the classical plasticity theory, set out in the Sections 1.1 and 1.2. When deformed under conditions T = const, reversible strains are observed for the stressed states $f(\sigma_{ij}, p_{ij}, \chi_i) < 0$.

The loading surface f, where χ_i are the parameters describing the loading history (structural parameters) under the condition of the von Mises maximum principle (1.16), is a plastic potential. The associated plastic flow law has the usual form (1.17). The kinetic equation for the structural parameters makes the system of equations closed.

The generalization of the flow theory proposed by the author of [58] is currently debatable.

The problem of correctly taking into account the viscosity of metals in describing their deformation, that is, the problem of an elastoviscoplastic body, despite the prescription of the formulation [46], does not have a consistent solution to date. The viscous flow, like the plastic flow, is irreversible, but unlike the latter, occurs at any stress, including a lower yield point. The physical mechanism of viscous deformation is controlled by diffusion [44, 45], so the problem of taking viscosity into account is especially important in deformation under high temperature and at moderate and low strain rates, for example, in deformation under conditions of the superplastic state of the material. Particularly relevant is the problem of taking viscosity into account in describing the non-stationary deformation processes and the creep process, as discussed in the Sections 1.2 and 1.3.

At present, viscosity is taken into account approximately when building deformation models with the use of mechanical (rheological) models of irreversible deformation mechanisms (see 1.3).

Returning to the author's proposal [58] concerning the solution of the first two problems on the path of generalization of the flow theory to the domain of finite strains, we note the following. Analysis of the solution shows that it, like the earlier ones proposed [57, 60], is not without its internal contradiction. The fact is that large irreversible strains of metals are inherently inhomogeneous in microvolumes. This is the cause of the occurrence of residual stresses [7]. It is known that these local stresses can reach the yield point [2]. Therefore, the author's assumption [58] that during the unloading process the changes in the components of the irreversible strain tensor are related only to the rigid displacement and rotation of microvolumes, is an approximation, but not the final solution of the problem. We also point out that the importance and magnitude of residual stresses are confirmed by the fact of constructing on their basis the plasticity theory with an allowance for kinematic hardening [19].

One should also note the absence of a more or less detailed experimental verification of the theory of developed strains.

In concluding this section, we note that the plasticity theory within the framework of a purely phenomenological approach continues to develop continuously. Both uniaxial [65] and triaxial models are beig developed and improved [59, 66].

The physical process of plastic deformation and fracture of metals is the object of research and another fundamental discipline – the physics of strength and plasticity. Chronologically, this science arose almost a century later than the mathematical theory of plasticity. The year of its birth is 1934, when G. Taylor, E. Orowan and M. Polanyi discovered dislocations. In recent decades, the physics of strength and plasticity has developed significantly [67–72].

The natural direction of the further development of the mathematical theory of plasticity is the unification of the ideas of mechanics and the physics of plastic deformation [73].

The number of works that develop this approach is continuously growing. Let us dwell in more detail on the results obtained, since this monograph is devoted to the development of this direction.

The analysis of the published works makes it possible to distinguish two approaches. The first is connected with the revision of the foundations of the mathematical theory of plasticity, taking into account the modern positions of the physics of strength and plasticity, but without going beyond the framework of a purely phenomenological description. This approach, the ancestor of which is considered to be N.K. Snitko [74], has been developing since the late 40s of the last century and led to the creation of a new version of the plasticity theory – the slip theory [75–79].

A great contribution to the development of this direction was made by V.A. Likhachev and co-authors. The results of these works he summarized in his last monograph [35], written in co-authorship with V.G. Malinin.

Let us briefly consider the main content of the approach using the example of [79], which is a development of the Batdorff–Budyansky model [75], and compare it with the approach of the

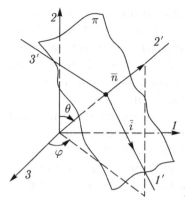

Fig. 1.18. Explanations in the text.

classical von Mises theory. The theory is based on physical concepts of plastic deformation as a shear process. In this case, the shears occur both in the body of grains at various sites, and along the boundaries of grains and fragments. From this it is concluded that when the representative element of a medium with a set of shears is deformed, it is divided into a set of small particles that deform elastically, and the deformation is ensured by the shears of the particles relative to each other. When deformed, the number of strong particles continuously increases without disrupting the continuity of the material. Increasing their number is necessary to continue the deformation. This explains the phenomenon of hardening.

It is assumed that the number of strong (deformable only elastically) elements in a representative elementary volume, and consequently the number of their shears, is very large, and the increments of discontinuous shifts are very small. Then the changes in this volume can be assumed to be the same as if it were an element of some effective medium. This assumption is justified from the point of view of the physical mechanism of irreversible deformation. It allows us to conclude that the element of the medium, which had the initial spherical shape under the action of the average displacement vector, will move somewhat, turn as a whole and take the shape of an ellipsoid in accordance with the 'mean' low strain tensor generated by the increment of the mean displacement vector. This assumption serves as the basis for the method of constructing the governing equations in the slip theory.

For the connection of the strain increment tensor with the stress increment tensor, the strain is represented as the sum of the elastic and plastic components.

In order to obtain the form of the strain increment tensor convenient for connection with the stress tensor, the entire set of shift areas in the volume element is sorted into groups in which a certain set of sites has a close orientation given by the unit vector of the normal **n**.

The increment of plastic deformation is divided into a sum of increments, each of which is caused by real shifts on planes with orientations **n**. It is assumed that the direction of the shift, given by the unit vector **i** (l_1, l_2, l_3) on all planes of the group, will be the same (Fig. 1.18). The orientation of the planes π is given by the spherical angles φ and θ, the totality of which is $\Omega = \Omega\,(\varphi,\theta)$. In this case, the projections of the vector **n** on the axis 1, 2, 3 will be

$$n_1 = \sin\theta\sin\varphi, n_2 = \cos\theta, n_3 = \sin\theta\cos\varphi.$$

Close orientations lie within the solid angle $d\Omega = \sin\theta\,d\theta\,d\varphi$.

Let $\Delta\gamma(\Omega)$ be the increment of the real shear density in the direction **i** of the group of shears along planes of close orientation. Then $\Delta\gamma(\Omega)d\Omega$ is the increment of the real shear in groups with orientations close to Ω.

The axes of the rectangular coordinate system x'_i are selected, as shown in Fig. 1.18. Then the increment in the shear density associated with the actual relative displacements $\delta\mathbf{u}\,d\Omega$ of strong particles in the direction **i** will be equal to

$$\Delta\gamma(\Omega) = \partial\delta u'_i / \partial x'_2 = 2\delta\varepsilon_{x'_1 x'_2}.$$

At the same time, the element will turn to an infinitesimal angle $\delta\omega_3\,d\Omega$, where

$$\delta\omega_3 = 1/2(\partial\delta u'_2/\partial x'_1 - \partial\delta u'_1/\partial x'_2).$$

In the laboratory coordinate system x_i this shear deformation corresponds to an increment of plastic strain equal to

$$\delta_\Omega\varepsilon_{ik} = 1/2\left(l_i n_k + l_k n_i\right)\Delta\gamma(\Omega)d\Omega.$$

The summation of strains caused by shears in all possible orientations Ω_0 gives

$$\Delta\varepsilon_{ik} = \int_{\Omega_0} \frac{n_i l_k + n_k l_i}{2}\Delta\gamma(\Omega)d\Omega.$$

The last equation is an analog of geometric relations in the plasticity theory.

A number of assumptions are made to obtain the defining relations, including assuming that the plastic shears in the planes of a given orientation begin when the tangential stresses τ on these planes exceed the yield point τ_y.

Omitting the intermediate arguments, we give the final form of the defining equations:

$$\Delta \varepsilon_{ik} = \left(A_{iklm}^{(e)} + A_{iklm}^{(p)} \right) \Delta \sigma_{im}, \quad (1.82)$$

$$A_{iklm}^{(e)} = (1+\nu) E^{-1} \delta_{il} \delta_{km} - \nu E^{-1} \delta_{ik} \delta_{lm}, \quad (1.83)$$

$$A_{iklm}^{(p)} = F_{ik} F_{lm}, F_{ik} = \int \sqrt{C} \, \frac{n_i \tau_k + n_k \tau_i}{2\tau} \left(\frac{\tau}{\tau_y} - 1 \right)^{1/2} d\Omega, \quad (1.84)$$

where $A_{iklm}^{(e)}$, $A_{iklm}^{(p)}$ are the coefficients of elastic and plastic compliance; E is the Young's modulus; ν is Poisson's ratio.

It is evident that the defining relations of the considered version of the slip theory connect at each calculation step the increment of the tensor of general strains with the increment of the stress tensor and have the form of the elasticity equations with variable coefficients of plastic compliance. In this case, the method of obtaining the generalized flow law differs fundamentally from the method in the flow theory.

The slip theory raises a number of questions, one of the main ones is the following. In mechanics, the most general formulation of the laws of motion is obtained using some variational principles that allow us to isolate the true (real) motion from the set of possible ones. The authors of [79] assert that the assumption that the direction of the increment of the shear on a system of planes with the orientation Ω coincides with the direction of the tangential stress vector $\tau(\Omega)$ in these planes adopted in the construction of the model is equivalent to the assumption that under plastic deformation each real shear occurs in the direction corresponding to the expenditure of the greatest possible work. This assumption is analogous to the maximum of the work of plastic deformation underlying the flow theory.

But then it turns out that the slip theory is equivalent to the so-called multi-surface flow theory, that is, each system of slip planes with a close orientation corresponds to its flow surface. This also agrees with the assumption made by the authors of [79] that there is a yield point τ_y in different sets of planes of close orientation.

In this case, for each set of slip systems, the components of the tensor $\Delta\sigma_{lm}$ in (1.82) must satisfy both the yield condition and the maximum of the plastic deformation work. This raises the question of the procedure for obtaining the defining equations (1.82)–(1.84). In any case, there is a contradiction, because there are a lot of flow surfaces in this case in a polycrystal, and it seems necessary to use statistical methods.

It should be noted that the multi-surface models of the flow theory are one of the directions of its development [59].

In constructing the model under consideration, it was assumed that small strains were assumed and that the model was verified in the range of small strains. The slip theory has not yet found application in calculating large strains that are observed in the processes of pressure metal working. In addition, like all phenomenological theories, it does not take into account the structure of the metal and its evolution in the process of deformation.

The second approach in the direction of the development of the theory of plasticity with the involvement of the provisions of the physics of strength and plasticity is associated with the direct use of the method of microscopic description of plastic deformation [80–84]. An analysis of these studies shows that the uniaxial determining equations of the plastic flow obtained in them, taking into account the physical concepts of the plastic deformation mechanisms and the method of its microdescription, contain a large number of parameters that are experimentally determined for each particular metal, which makes them non-universal. We do not consider triaxial (generalized) defining equations, which are the apex of the plasticity theory.

This brief review of the current state of the mathematical theory of plasticity allows us to formulate a scientific problem, the solution of which will determine the development of the theory. This problem includes the following tasks.

1. The problem of describing large deformations, which are typical, for example, for processes of pressure metal working. In these processes, the accumulated plastic strain intensities (Udquist parameter) reach $\varepsilon = 2.0$ or more [2]. For large strains, in addition to the hardening process, softening processes and, often, deformation diagrams $\sigma(\varepsilon)$ have time to proceed, where σ is the stress intensity (effective stress), ε is the strain rate, are non-monotonic at elevated temperatures (so-called 'falling' deformation diagrams). Such

plastic behaviour is excluded from consideration by the classical mathematical theory of plasticity, since it contradicts the Drucker postulate [7, 8].

Large strains, as a rule, take place under the conditions of complex loading and non-monotonic deformation. From the thermodynamic point of view, plastic deformation and viscous destruction are a single non-equilibrium (irreversible) physical process, therefore the current characteristics of the process depend not only on the current thermomechanical loading parameters, but also on the history of their change. A consistent account of the loading history is one of the unsolved problems of the classical theory of plasticity [1, 2, 5].

2. The problem of elastic–viscoplasticity and, accordingly, creep is also associated with the process of softening and taking into account the loading history [7, 46, 59].

3. The problem of accounting for the structure of the metal and its evolution in the processes of irreversible strains and viscous fracture. Within the framework of a purely phenomenological approach, this problem can not be solved, since the material model in this approach is a continuous structureless medium.

The urgency of the formulated scientific problem is continuously increasing. This is due to the fact that the mathematical theory of plasticity is the fundamental basis of applied sciences, including the theory of pressure metal working, which formed into an independent discipline in the 20–30s of the twentieth century. The theory of pressure metal working develops in two directions: mechanical and mathematical (based on the mathematical theory of plasticity); physical with elements of physical chemistry (on the basis of the physics of strength and plasticity) [2].

It has been repeatedly asserted that the most developed in theory is the mechanico–mathematical direction [2, 85–88], which corresponds to reality. All calculations and studies of technological processes of pressure metal working are carried out on the basis of the applied theory of plasticity, and the references given above to textbooks and monographs on the theory of pressure metal working are devoted to the exposition of its foundations.

However, from the above brief review of the state of the mathematical theory of plasticity, it follows that at the present time the theory lags behind the demands of practice (modern technology of pressure metal working).

Multi-stage cold bulk forging (MSBF) is one of the effective technologies of pressure metal working and is increasingly used. Its

Fundamentals of Mechanics of Strength and Plasticity

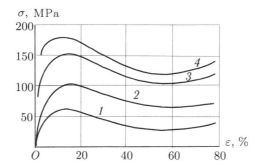

Fig. 1.19. The deformation diagrams of VT9 titanium alloy obtained by upsetting at $T = 850°C$: $1 - \dot{\varepsilon} = 4 \cdot 10^{-4}$ s^{-1}; $2 - 1.6 \cdot 10^{-3}$ s^{-1}; $3 - 8 \cdot 10^{-3}$ s^{-1}; $4 - 4 \cdot 10^{-2}$ s^{-1}.

undoubted merits are as follows: the coefficient of metal utilization is $K = 0.90$–0.98; plastic forming parts are made of non-ferrous alloys and steels with a roughness parameter Ra of 0.8 to 0.08 µm; the accuracy of the dimensions is 10–8 on the ISO scale; the productivity at stamping on cold-pressing machines is 5–10 times higher than the productivity when shaping similar parts by cutting on modern machines [89]. The technology is being developed on the basis of the production experience, since the plasticity theory does not allow for the deformation anisotropy of the properties of the material being processed and, correspondingly, the complex loading that arises during cold deformation.

Hot volume forging is the most common type of pressure metal working [90], it develops in the direction of increasing the accuracy of stamped forgings, reducing material and energy consumption, processing new high-alloy and low-deformability alloys with special functional properties. The mathematical theory of plasticity provides a very approximate calculation of the stress–strain state of the billet under hot volume forging, since it does not take into account the loading history associated with deformation, strain rate and temperature dependences on the deformation time, i.e., the non-stationary nature of the process. This is due to the fact that the problem of an elastoviscoplastic body has not been solved for many years in the framework of the theory of plasticity – a purely phenomenological approach [59, 91, 46].

Isothermal stamping, including under the superplasticity conditions of the processed material [91–93], gradually finds its technological niche. However, calculations and studies of these technological processes on the basis of the classical theory of plasticity are problematic. The matter is that materials under conditions of superplastic deformation have the so-called falling deformation diagram. Many high-alloy steels and alloys with hot deformation also have such a diagram (Fig. 1.19). To these materials, as noted in section 1.1, the Drucker postulate is not applicable, from which the known maximum principles underlying the theory of plasticity follow.

Solving these problems within the framework of the phenomenological approach encounters great difficulties.

The urgency of the formulated problem has substantially increased in connection with the intensive development of the technology of plastic structure formation of metals – the production of the ultrafine-grained state of structural alloys by the method of large plastic deformation, which have an increased level of mechanical properties compared with the coarse-grained state [34, 94]. The theory of this process of pressure metal working should give a method for determining the characteristics of a stress–strain state and predicting the evolution of structure characteristics during plastic processing of blanks. In the processes of plastic structure formation, a cold or warm deformation is cyclic or close to it, while the accumulated strain intensity reaches a level of 10–12 or more.

At present, it is generally accepted that plastic deformation and the destruction of metals are a single kinetic multistage physical process [72, 73]. As a consequence, in recent years the so-called coupled plasticity models have been developed. Attempts are made to take into account the effect of deformation damage of the metal, which occurs during deformation, on the deformation itself. These models are considered in detail in [80]. In this area of research, which is under development, two approaches are also used: phenomenological and physical. We only note that the physico–mathematical theory of a single process of irreversible deformation and viscous destruction of a metal, presented in this monograph, is a coupled theory.

1.4.2. Creep theory

As noted in section 1.3, the classical phenomenological theory of creep has received a significant impetus in its development with the publication of the paper [41]. In it, Yu.N. Rabotnov proposed for a uniaxial stress state a system of equations consisting of an equation called the *creep state equation* and differential kinetic equations describing, from a phenomenological point of view, a change in the parameters introduced into consideration that were called structural and interpreted as measures of material damage. In this case, the

Fundamentals of Mechanics of Strength and Plasticity

uniaxial equation of state describes the dependence of the creep rate of a structurally stable material on the stress, temperature, and structural state of the material at each instant of time. This theory was recognized in world science and was called *kinetic*.

Considering the current trends in the development of the creep theory, let us first of all look at a fairly developed energy version of the kinetic theory of creep and long-term strength of metals [95, 96], which describes creep under conditions of a complex stress state.

Let us analyze the hypotheses put by the authors in the basis of the theory [96].

1. The material is considered incompressible until the moment of destruction, i.e.

$$\dot{\varepsilon}_{ij}^c \delta_{ij} = 0. \qquad (1.85)$$

For conventional industrial structural materials, this statement is fully justified. It allows us to identify the tensor of creep strain rates $\dot{\varepsilon}_{ij}$ with its deviator.

2. The proportionality of deviators $\dot{\varepsilon}_{ij}^c$ and σ_{ij}^c is accepted. This is equal to the axiom of existence of the creep potential

$$f^c = \sqrt{\frac{3}{2} s_{ij}^c s_{ij}^c} = \sigma,$$

where σ is the stress intensity; s_{ij}^c is the deviator of the stress tensor σ_{ij}^c, and the adoption of the associated flow law in the form (1.60)

$$\dot{\varepsilon}_{ij}^c = \lambda \frac{\partial f^c}{\partial \sigma_{ij}^c}, \qquad (1.86)$$

where λ is a non-negative function.

Multiplying (1.86) by σ_{ij}^c, carrying out the summation and taking into account that f^c is a homogeneous stress function of the first degree, we obtain

$$\lambda = w/\sigma^c, \qquad (1.87)$$

where $w = \sigma_{ij}^c \cdot \dot{\varepsilon}_{ij}^c$ is the specific dissipation power.

The second hypothesis shows that when constructing the theory of deformation, the authors used the classical method described in the Sections 1.1 and 1.2, hence the creep theory absorbed all the shortcomings mentioned in 1.2.

3. It is assumed that there is a functional relationship between the intensities of the stress tensors σ^c, the creep strain ε^c, and the strain rates $\dot{\varepsilon}^c$. Consequently, under a uniaxial stress state we have a well-known dependence of the hardening theory and inevitably accepts the independence of this function from the form of the stress state – the hypothesis of a single curve, which, as is known, is not always satisfied.

In the present version of the creep model, in contrast to the traditional approach, the existence of a functional relationship between the stress intensity, the specific dissipation work A and the specific dissipation power W, is postulated, with $dA = W\,dt$. As a measure of the creep process, specific work is used instead of the intensity of deformation, and as a measure of the intensity of the process, the specific power of scattering instead of the traditional rate of creep strain. What does it give?

The transition to new measures eliminates the lack of technical creep theories, which is associated with the non-equivalence of the concepts of the Udquist parameter

$$q^c = \int_0^t \dot{\varepsilon}^c dt$$

and the intensity of the accumulated creep strain

$$\varepsilon^c = \int_0^t \dot{\varepsilon}^c \cos(\varphi - \psi) dt,$$

where φ and ψ are the angles of the tensors $\dot{\varepsilon}^c_{ij}$ and ε^c_{ij}. These measures are equivalent for $\varphi = \psi$, which is valid only for the stationary stress state and proportional loading. With new scalar deformation measures, for any loading path $\sigma^c_{ij}(t)$ always $dA/dt = W$. Therefore, the energy version of the creep theory of creep should be considered as its generalization to a complex stress state.

4. The concept of the mechanical equation of state proposed by Yu.N. Rabotnov is generalized, which, with the application of new measures for the case of a complex stress state, is as follows: "The intensity of the creep process of a structurally stable material at each instant of time is a function of the stress intensity, temperature T, and structural state of the material:

$$W = W\left(\sigma^c, T, A, \omega_1, \omega_2, \ldots, \omega_k\right) \tag{1.88}$$

(ω_1, ω_2, ..., ω_k are the parameters describing from the phenomenological standpoint the change in the structure of the material, due to the accumulation in it damage)".

In this approach, expression (1.88) is an analog of the loading function f^c. At this point, obviously, there is a shortage of all the classical phenomenological theories of irreversible strains associated with the phrase 'structurally stable material'. Irreversible deformation occurs due to a change in the structure of the material.

5. Following the kinetic theory of creep, the parameters ω_k are identified with the damage measures. For simplicity, one parameter is taken, and the kinetic equation of its variation:

$$d\omega/dt = F(\sigma^c, T, A, \omega). \tag{1.89}$$

For $t = 0$ we have: $\omega = 0$; at $t = t^*$ $\omega = \omega^* = 1$, where $\omega^* = 1$ is the critical value of damage. Consequently, $t = t^* = \psi(\sigma_*^c)$ and σ_*^c is the criterion for the long-term strength of the material.

The parameter ω is phenomenological in nature. Equation (1.89) for it is formulated as a hypothesis, since for a phenomenological parameter it is impossible to formulate a sufficiently rigorous physical model.

The energy version of the kinetic theory of creep and long-term strength has been experimentally substantiated by the example of a number of metals in a wide range of loads and temperatures under the conditions of simple and complex loading [97], and also has examples of applications for the development of a specific technology of plastic shaping [98].

The theory is a further development of the problem of creep of metals. However, it retains a number of shortcomings inherent in a purely phenomenological approach in the cognition of the processes and phenomena of nature.

The energy version of the theory of creep and long-term strength continues to be developed by students and followers of O.V. Sosnin [99], including attempts to describe the creep of anisotropic materials [100].

Apparently, the consequence of deep internal unity of processes of active plastic deformation and creep is the fact of applying the same approaches to describing and further developing these irreversible strains. Thus, a non-linear creep model based on the slip concept was proposed in [101], and a stochastic model of non-isothermal creep and long-term strength of materials is described in [102]. Without dwelling on an analysis of these approaches to the development of creep theory, we note the following. All new models that use the phenomenological approach to the study of the physical process are directed initially (at the stage of formulating postulates and hypotheses) to solve specific deficiencies in the theory. Therefore, eliminating one drawback of the theory, new hypotheses, not being, as a rule, sufficiently general, introduce new ones.

In conclusion, we note that the problem of the creep and long-term strength of metals is, as already noted, the problem of a viscoplastic body, which continues to be relevant.

2

Fundamentals of the phenomenological theory of fracture and fracture criteria of metals at high plastic strains

2.1. Basic concepts, assumptions and equations of the phenomenological theory of the fracture of metals

The description of the physical process of the fracture of metals, caused both by external actions of non-termomechanical nature (chemical reactions, irradiation) and thermomechanical action, is based on the concept of fracture as a loss of the material's ability to resist deformation due to violation of internal intermolecular and interatomic bonds [103]. One of the basic concepts in the theory of fracture is the notion of material damage, which is understood as a violation of the continuity of its structure due to external influences. Damage is a parameter of the state of the material when considering the process of its fracture. The damage of *brittle* elastic materials and the damage of *ductile* elastoplastic materials beyond the elastic limit is considered.

The process of accumulation by the material of damage and fracture under the influence of external thermomechanical action (*deformation damage*), like any other process, can be studied within the framework of two approaches – phenomenological and statistical. We will begin, as it is expected in science, from the phenomenology of the phenomenon.

Forecasting the deformation damage and fracture of the workpiece during pressure treatment, as well as any other structure during its service, is a necessary condition for obtaining quality parts and their blanks, as well as determining the durability and residual life of machine parts and structures. Therefore, the level of development of the theory of *deformability* (*ductile fracture of metals in shaping and forming operations and pressure metal working processes, which are characterized by large strains and complex loading*) also determines the level of development of pressure metal technology and methods of design calculations.

Processing of metals by forging and stamping is based on the plastic redistribution of the volumes of the initial billet. Plasticity means the property of solids to irreversibly change the shape under the action of external forces without a macroscopic discontinuity. The fact that this property depends on the stress state scheme and, in general, on the law of continuous structure change, generally does not allow us to introduce a sufficiently strict measure of plasticity. The degree of shear deformation (Λ_*) accumulated by the macroscopic sample at the instant of macroscopic discontinuity appearance is considered to be a conditional measure of plasticity, i.e., T, H, K, μ_σ = const, where $H = \sqrt{3}\dot{\varepsilon}$ is the rate of shear deformation, $K = \sigma_0/\tau$ is the rigidity index of the stressed state, $\sigma_0 = 1/3\sigma_{ii}$ is the average normal stress, τ is the shear stress, $\mu_\sigma = (2\sigma_2-\sigma_1-\sigma_3)/(\sigma_1-\sigma_3)$ is the stress state indicator (the Nadai–Lode parameter). From the constancy of the thermomechanical parameters, the need for uniformity of deformation of the macrosample follows. Hence, the complexity of the experimental determination of this conditional plasticity measure becomes clear. In addition, this measure can only be used for comparative studies of the plastic properties of different materials under identical conditions.

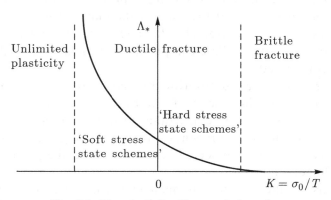

Fig. 2.1. The plasticity diagram (scheme).

Fundamentals of the Phenomenological Theory of Fracture

The plasticity of a concrete material, as described above, is described by the *plasticity diagram* $\Lambda_*(K)$ – the dependence of the limiting (up to fracture) shear deformation on the rigidity index of the stress state (Fig. 2.1), which is constructed from the test of special specimens by tension or torsion in a liquid high pressure [2, 104].

The dependence of the limiting plastic deformation on the loading history (for T, H, K, μ_σ = var) determined the introduction of the concept of deformability. In fact, it is a certain concretization of the concept of plasticity. Under deformability is meant the ability material to be plastically deformed in real technological processes and plastic shaping operations, i.e., under conditions of varying thermomechanical parameters. A measure of deformability is the degree of shear deformation or the strain rate $\varepsilon_* = \Lambda_*/\sqrt{3}$ accumulated by an elementary volume (material particle) at the time of macroscopic discontinuity, when the dependence of the thermomechanical parameters on the deformation time is determined for a particular test method or the technological operation of pressure metal working. A measure of deformability can also be the deformation damage accumulated by an elementary volume, described by some parameter. At the moment of fracture, the damage reaches a critical value. From this definition it follows that the theory of deformability is *the theory of ductile fracture of materials under developed plastic strains*.

The development of ideas about the fracture of metals as a physical process has a long history and is reflected in many reviews and monographs [72, 105–108]. The most important achievement in the problem of fracture is the rejection of the idea of it as a critical phenomenon and recognition of the fracture as a kinetic, probabilistic, multistage and multiscale process of the formation and accumulation of deformation damage [71–73, 109].

Initially, the moment of fracture was associated with the attainment of stresses, strains, or specific work of plastic deformation of the limiting (critical) values: $\sigma = \sigma_*$, $\varepsilon = \varepsilon_*$, $A = A_*$, and a *force criterion of fracture* was formulated. According to this criterion, the local stresses at the origin of the microcrack or at its tip should reach the theoretical shear strength $\tau_{theor} \approx 0.2G$, i.e.,

$$\tau_{loc} = \tau_{theor}. \tag{2.1}$$

Taking into account the fact that real bodies are destroyed at stresses one or two orders of magnitude less than τ_{theor}, Griffiths developed the concept of the fracture of absolutely brittle bodies as a first-order phase transition, in which the new phase is the discontinuity (microcrack). At the same time, he postulated the proposition that the nuclei of the new phase in the form of submicrocracks are initially present in the material [72].

For the potential energy of a disc-shaped microcrack of size h in a body under the action of a tensile stress σ,

$$w = 2\pi\gamma h^2 - 2/3\,\pi(\sigma^2/E)h^3, \qquad (2.2)$$

where $\gamma = Ga/8$ is the specific surface energy; a is the parameter of the crystal lattice; E is the Young's modulus.

The equilibrium size of the microcrack from the condition $\partial w/\partial h\big|_{h=h_{\text{Gr}}} = 0$ is equal to

$$h_{\text{Gr}} = 2\gamma E/\sigma^2. \qquad (2.3)$$

It follows from (2.2) that

$$\partial^2 w/\partial h^2\big|_{h=h_{\text{Gr}}} = -4\pi\gamma < 0,$$

which indicates the instability of the equilibrium for $h = h_{\text{Gr}}$. At $h = h_{\text{Gr}}$, the growth of the microcrack is energetically favourable, for $h < h_{\text{Gr}}$ it is advantageous to close it (healing). The characteristic size of the microcrack h_{Gr} is called the *size of the Griffith crack*.

In plastic materials, such as metals, the growth of local stresses at the tip of the microcrack inevitably leads to the appearance of local plastic deformation (plastic zone) and blunting of its tip before the stress σ_1 reaches the value $\sigma_{\text{loc}} = \sigma_{\text{theor}}$. This causes a decrease in the level of local stresses up to

$$\sigma_{\text{loc}} \approx \sigma_{\text{theor}}\sqrt{a/\xi}, \qquad (2.4)$$

where ξ is the radius of curvature of the tip of the microcrack after the local plastic deformation.

Consequently, in the general case ($\xi \gg a$), there exists a local stress range $\sigma_{\text{theor}} > \sigma_{\text{loc}} > \sigma_w$, where σ_w is the stress at which the energy criterion for microcrack instability is fulfilled

$$\partial w/\partial h \leqslant 0, \qquad (2.5)$$

under which the rupture of interatomic bonds is thermodynamically

Fundamentals of the Phenomenological Theory of Fracture 61

advantageous, but can not be realized because of the failure of the force criterion (2.1) [72].

Hence the following two fundamental provisions result from this:

1) the energy criterion (2.5) is necessary, and the power criterion (2.1) is a sufficient condition for the propagation of a microcrack;

2) in metals for which relation (2.4) is valid, the force criterion can be satisfied only due to the action of local thermal fluctuations that can overcome the potential barrier.

From the second position follows the *kinetic concept of fracture* proposed by S.N. Zhurkov [109].

Based on the concept of fracture as a process of consequent accumulation of damage, various versions of theoretical models were proposed for its description [110–117]. Scalar and tensor quantities are proposed as measures of damage. In most cases, the measure of damage is introduced as follows [116]. Fracture occurs when the accepted parameter of damage q reaches the critical value q_*. Then the damage measure $\psi = q/q_*$, or, in the differential form, $d\psi = dq/q_*$, and also q and q_* are functions of some loading parameters. When using the hypothesis of linear summation of damage, the failure condition will look as follows:

$$\psi = \int \frac{dq}{q_*} = 1. \tag{2.6}$$

Within the framework of the phenomenological approach and taking into account the kinetic concept of fracture, the phenomenological theory of deformability proposed by Kolmogorov and Bogatov [2, 104] is the most developed model of accumulation of deformation damage and ductile fracture of metals under large plastic strains. This theory has a number of examples of application in the analysis and design of metal pressure working processes. Let us briefly consider its main provisions and equations.

V.L. Kolmogorov, relying on the ideas of Yu.N. Rabotnov [41, 118] and assuming that ψ is a sufficiently smooth deformation function, formulated the following hypothesis. *The elementary deformation damage of a material particle $d\psi$ under monotonic deformation is directly proportional to the elementary increment of plastic deformation $d\Lambda$ and inversely proportional to the plasticity of the material Λ_*:*

$$d\psi = d\Lambda/\Lambda_*. \tag{2.7}$$

In this case, taking into account the above-described method for determining the value of Λ_*, the particle deforms at constant values of the thermomechanical parameters.

Here it is worth noting the important fact that the phenomenological and statistical or microstructural approaches to the study of the physical process differ not only in axiomatics and methods, but also predetermine different types of thinking. The mechanical engineer represents the elementary deformation damage in the metal in the form (2.7), and the physicist, as we shall see below, in the form of a dislocation submicrocrack in the crystal lattice.

The parameter ψ is scalar and, hence, additive and invariant. The damage for the entire stage of monotonic deformation with $H = \text{const}$ is determined by taking into account $d\Lambda = H\,dt$, where t is the time, as

$$\psi = \int_0^t \frac{H}{\Lambda_*}\,dt. \qquad (2.8)$$

Experience shows that the limiting plasticity Λ_*, like the resistance of metals to plastic deformation, is a functional of the form $\Lambda_* = \Lambda_*[T(t), K(t), \mu_\sigma(t), H(t), t]$. Therefore, the generalization of (2.7) for a particular region of monotonic deformation yields the kinetic equation [2]

$$d\psi/dt = H(t)\big/\Lambda_*\big[T(t), K(t), \mu_\sigma(t), H(t), t\big]. \qquad (2.9)$$

The function ψ is normalized. According to (2.8), with $Ht = \Lambda = \Lambda_*$, the damage is 1.0, and this is interpreted as the appearance of a macroscopic discontinuity (macrocrack visible to the naked eye) in the elementary volume under consideration. Deformation without fracture occurs under the condition $\psi < 1.0$. Therefore, ψ can also be interpreted as the *degree of use of the plasticity resource of a material* under deformation under the given conditions, and as the *probability of macrofracture*.

The use of (2.8) and (2.9), for example, to determine the damage to a workpiece being processed by pressure, presupposes a preliminary statement and solution for the considered technological operation of pressure metal working of the boundary value problem of the theory of plasticity with a complete determination of the stress–strain state (SSS) of the workpiece from the beginning of the deformation to its end. Knowledge of SSS allows us to determine the functions $K(t)$, $\mu_\sigma(t)$ and $H(t)$. However, the functional Λ_* in the denominator of the right-hand side of (2.9), also called the limiting plastic deformation in the literature [106], remains unknown.

A.A. Bogatov formulated, within the framework of the approach and concepts of the model of V.L. Kolmogorov, a non-linear model of damage cumulation. He reasoned as follows [104].

Assume that at the initial time $t = 0$, the damage intensity as a function of the degree of shear deformation with other arguments unchanged is zero: $d\omega/d\Lambda = 0$[1], and at the moment of macrofracture, at $t = t_*$, is inversely proportional to the material plasticity $d\omega/d\Lambda = a/\Lambda_*$. At the current time $0 < t < t_*$, the rate of damage cumulation depends on the value of ω accumulated up to this moment. In this case, the damage to the elementary volume at the stage of monotonic deformation is defined as

$$\omega = \int_0^{\Lambda_1} \frac{a\Lambda^{a-1}}{\Lambda_*^a} d\Lambda, \qquad (2.10)$$

where

$$a = a(K,\mu_\sigma,T,H,c_f,x_f), \Lambda_* = \Lambda_*(K,\mu_\sigma,T,H,c_f,x_f); \qquad (2.11)$$

c_f and x_f are the characteristics of the structure and chemical composition of the metal.

The authors of [2, 104] call the functions (2.11) the defining relations of the model of *ductile fracture of metals under developed plastic strains*. They are determined experimentally when the special specimens are deformed by tension or torsion in a high-pressure liquid at special installations with varying parameters.

On the basis of the model (2.10), (2.11) the deformability theory is developed with reference to the conditions of monotonic and alternating deformations. When calculating the process of plastic shaping of a billet under monotonic deformation conditions, on the basis of the results of the SSS study, it is divided into n stages, in each of which the degree of deformation has the value Λ_i, and the thermomechanical parameters are constant, but change discontinuously when passing from i to $(i + 1)$th stage. In this case, on the basis of (2.10), for n deformation stages, the following recurrence formula holds

---

1) Here we retain the notation of damage ω, adopted by the author [104].

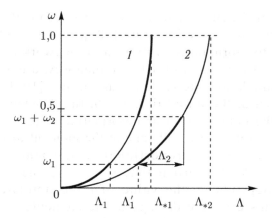

Fig. 2.2. Scheme of damage cumulation under monotonic deformation in two stages.

$$\omega = \left\langle \left\{ \left[\left(\frac{\Lambda_1}{\Lambda_{*1}} \right)^{a_1/a_2} + \frac{\Lambda_2}{\Lambda_{*2}} \right]^{a_2/a_3} + \ldots + \frac{\Lambda_i}{\Lambda_{*i}} \right\}^{a_i/a_{i+1}} + \ldots + \frac{\Lambda_n}{\Lambda_{*n}} \right\rangle^{a_n}. \quad (2.12)$$

Let us explain it with the example of calculating two stages (Fig. 2.2) [104]. Lines *1* and *2* reflect the damage cumulation in an elementary volume when it is deformed from the initial state, when $\omega_0 = 0$ is assumed, to macrofracture ($\omega = 1.0$), with constant thermomechanical parameters corresponding to stage 1 and stage 2 when forming the workpiece.

In stage 1, the material particle undergoes a deformation Λ_1 and, in accordance with (2.10), accumulates the damage $\omega_1 = (\Lambda_1/\Lambda_{*1})^{a_1}$. In step 2, the particle deforms at the new values of the thermomechanical parameters, but at the beginning of the stage the damage is equal to ω_1 and, consequently, $\omega_1 = (\Lambda_1'/\Lambda_{*1})^{a_2}$, where Λ_1' is the degree of deformation at which the particle accumulates a damage equal to ω_1, deforming initially at the thermomechanical parameters of stage 2. From the two equations for ω_1 we obtain

$$\Lambda_1' = \Lambda_{*2} (\Lambda_1/\Lambda_{*1})^{a_1/a_2}.$$

From the scheme in Fig. 2.2 it follows that the damage to the particle in phase 2 is

Fundamentals of the Phenomenological Theory of Fracture

$$\omega_2 = \left[(\Lambda'_1 + \Lambda_2)/\Lambda_{*2}\right]^{a_2} - (\Lambda'_1/\Lambda_{*2})^{a_2} = [(\Lambda_1/\Lambda_{*1})^{a_1/a_2} + \Lambda_2/\Lambda_{*2}]^{a_2} - \omega_1.$$

The total damage during monotonic deformation in two stages is

$$\omega = \omega_1 + \omega_2 = [(\Lambda_1/\Lambda_{*1})^{a_1/a_2} + \Lambda_2/\Lambda_{*2}]^{a_2}.$$

For the defining relations (2.11), on the basis of a generalization of the experimental data the following approximate dependences were proposed

$$\Lambda_* = \chi \exp\left(\lambda \frac{\sigma_0}{\tau}\right), \qquad a = a_0^{1+b\sigma_0/\tau}, \tag{2.13}$$

where χ, λ, a_0, b are the coefficients that for each metal (alloy) are determined by the results for the experiment model.

At first glance, it follows from the example considered that the recurrence formula (2.12) ensures that the calculation of ω takes into account the loading history due to the dependence $K(t)$. However, a more detailed analysis of the model shows that this is not quite so. The fact is that the analyzed model does not take into account the decrease in damage due to the healing of microdefects during deformation.

Turning again to the example of calculation of ω analyzed above for two stages of monotonic deformation, one can note the following. In step 2, the deformation occurs with a softer stress-strain scheme state ($\Lambda_{*2} > \Lambda_{*1}$), that is, $K_1 > K_2$ (see Fig. 2.2). Therefore, at the beginning of stage 2, the damage accumulated in stage 1 should decrease due to healing of microdefects under the action of compressive stresses. The fraction of these stresses in the stressed state is greater than in stage 1. Therefore, as the deformation develops, ω will begin to increase again, but not in accordance with curve 2 (Fig. 2.2). Obviously, with an increase in the number of calculated steps, the error in determining ω will increase.

This remark is also true for the phenomenological model of deformability for alternating deformation. Thus, it was established in [117] that when the loading conditions change, a transient process takes place during which the accumulation of damage to the changed conditions takes place. If the change in the loading conditions occurs in a favourable direction (for the manifestation of high deformability of a metal), a partial decrease in the damage (healing of microdefects) is observed.

Work [117] is one of the last doctoral dissertations devoted to the further development of the phenomenological theory of deformability from the point of view of the loading history. Without dwelling on the description and analysis of its results, which undoubtedly have significant scientific significance, we note the following. In our opinion, the main and most important result of this work is convincing evidence of the presence of significant difficulties on the way to an adequate description of the deformability of metals within the framework of a purely phenomenological approach. It can even be concluded that this approach to solving the problem, apparently, completely exhausted itself. This is evidenced by the presence of more than two dozen phenomenological coefficients in the equations of the deformation damage model developed in [117], which must be determined from the results of laborious and methodically complex experiments requiring special equipment.

It should be noted, however, that the scientific significance of the phenomenological theory of metal deformability can not be overestimated. With its development, a clear understanding of the problems and tasks on the way to the theoretical description of the ductile fracture of metals under developed plastic strains is achieved. The basic concepts and criteria that are theoretically and experimentally substantiated are formulated. Among them, it is necessary to note the criteria for microdamage ω_* and ω_{**}. If the damage is $\omega < \omega_* = 0.25$, microdefects are healed during subsequent recrystallization annealing. The presence in the metal of damage $\omega > \omega_{**} \approx 0.62$ leads to a decrease in the strength characteristics in comparison with the parent metal.

The phenomenological theory of deformability in the framework of its approach and initial assumptions is the most complete and has a large experimental material, including numerous plasticity diagrams constructed for different alloys and various thermomechanical conditions of deformation. This forms a reliable basis for the formulation and testing of new ideas and the further development of the theory of deformability.

2.2. Criteria of ductile fracture of metals

Ductile fracture of metals occurs both in the process of active plastic deformation, for example, in the pressure treatment of metal [117], and as a result of creep [52, 53, 119] and metal fatigue [120, 121]. The process of fracture depends on the method of deformation. As

Fundamentals of the Phenomenological Theory of Fracture 67

before, we will be mainly interested in ductile fracture in the process of active plastic deformation (short-term strength). Nevertheless, we note that when predicting the fracture in the three noted cases, two approaches are used: kinetic and criterial. The content of the first approach was considered in section 2.1 with the example of the phenomenological theory of deformability. Let's briefly consider the criteria for ductile fracture of metals.

The basic concept under the criterial approach is the notion of *equivalent stress* [122]. As such, four basic combinations of principal stresses σ_1, σ_2 and σ_3 are considered: the maximum principal stress σ^*, for which, as a rule, a tensile stress is taken; σ is the stress intensity; σ_0 is the average normal stress and half-sum $(\sigma^* - \sigma)/2$.

All the fracture criteria are phenomenological, that is, they lack the characteristics of the structure in an explicit form. They relate the damage of the material ψ obtained during the deformation process to the effective stress and the increment in strain intensity $d\varepsilon$. It is believed that the microfracture corresponds to the equality of ψ to some limiting value ψ^*.

Damage criteria used for active plastic deformation are not related to the defining relationships, that is, they are said to be disconnected. And the process of fracture is regarded as an independent process. Since physically this is not true, the fracture criteria are of only practical importance.

Let's consider some criteria that are built into the DEFORM software complex of mathematical modelling of the processes of pressure metal working [123]. They are subdivided into energy, deformation and force.

The idea of the Freudenthal energy criterion often applied abroad of the type

$$\psi = \int_0^{\varepsilon_*} \sigma d\varepsilon, \qquad (2.14)$$

where ε_* is the intensity of the deformation accumulated at the time of fracture, is related to the notion that the initiation of macroscopic fracture is caused by the discontinuity of the collective of interatomic bonds. Consequently, a certain potential energy must be stored by the time of fracture in this set of connections [124, 125].

Another energy criterion, which has examples of applications in the analysis of pressure metal working processes, is the Cockroft–Latham criterion [126]

$$\psi = \int_0^{\varepsilon_*} \frac{\sigma^*}{\sigma} d\varepsilon. \tag{2.15}$$

To use (2.15) in the calculations, an experimental determination is necessary for the material in question and the method of deformation of the limiting value ψ^*. For $\psi < \psi^*$ there is no macrofracture.

The energy criteria also include the Oyane, Ayada and Brozzo criteria, which have the corresponding form

$$\psi = \int_0^{\varepsilon_*} \left[1 + \frac{1}{a_0}\frac{\sigma_0}{\sigma}\right] d\varepsilon, \tag{2.16}$$

$$\psi = \int_0^{\varepsilon_*} \frac{\sigma_0}{\sigma} d\varepsilon, \tag{2.17}$$

$$\psi = \int_0^{\varepsilon_*} \frac{2\sigma^*}{3(\sigma^* - \sigma)} d\varepsilon. \tag{2.18}$$

For completeness of the presentation, we give the formulations of the Rice & Tracy, Osakada deformation criteria,

$$\psi = \int_0^{\varepsilon_*} \exp\left(\frac{\alpha \sigma'}{\sigma}\right) d\varepsilon, \tag{2.19}$$

$$\psi = \int_0^{\varepsilon_*} \exp(\varepsilon + a\sigma_0 - b) d\varepsilon, \tag{2.20}$$

and the Zhao & Kuhn force criterion

$$\psi = \sigma^*/\sigma. \tag{2.21}$$

An analysis of the studies in which the correspondence between the given fracture criteria and the experimental data was carried out [127–132] shows that ductile fracture depends on a large number of factors. The main are: intensity of stress σ; loading scheme; the Nadai–Lode parameter v_σ; the rigidity index of the stress state scheme $K = \sigma_0/\tau$. Moreover, each of the phenomenological criteria of fracture is adequate to experience only in certain ranges of variation of the factors noted above [127], i.e., there is no universal criterion.

Fundamentals of the Phenomenological Theory of Fracture 69

The application of the above criteria is associated with a large volume of the basic experiments for determining the material constants [131], an increasing their adequacy requires an increase in the scope of the experiment [129].

The proposed fracture criteria do not consistently take into account the physical mechanism of healing of deformation damage carriers (microcracks and microvoids) under the action of a positive hydrostatic pressure $p = -\sigma_0$ [133]. Therefore, the existing criteria do not give a prediction of fracture, corresponding to experiment, under deformation schemes in which σ_0 and, respectively, $K < 0$ [127].

In this regard, in the documentation of the DEFORM program complex it is recommended to use the fracture criteria for the estimated calculations and not to consider the results as unambiguous.

According to modern concepts, irreversible deformation and ductile fracture of metals are a single physical process. In this connection, the disconnected criteria of fracture considered, having some practical significance, do not have a scientific value. Awareness of this fact has led to an increasing number of works, in which attempts are made to construct a unified theory of deformation and ductile fracture [80], where, as in the description of irreversible deformation, two approaches are realized – phenomenological and physical. In the latter case, the model is based on certain micromechanisms of irreversible deformation and ductile fracture.

2.3. Modern approaches to the development of the theory of ductile fracture and the formulation of a scientific problem

Perhaps the most famous model of the damage to an elastoplastic material, in the construction of which physical representations were used, is the mesomechanical[1] Garzon–Tvergaard–Nidelman model (GTN-model) [134–136].

According to the GTN model in the elastic region, the material obeys Hooke's law

$$\sigma_{ij} = D_{ijkl}\varepsilon_{ij}^e, \qquad (2.22)$$

where D_{ijkl} are the tensors of elastic constants.

A scalar – porosity – is taken as a measure of damage. The strain rate is the sum of the rates of elastic and plastic strains:

---
1) Mesomechanics – one of the names of the section of the mechanics of a deformable solid in which irreversible deformation is described taking into account the physical notions of micromechanisms of its development and subsequent fracture.

$$\dot{\varepsilon}_{ij} = \dot{\varepsilon}_{ij}^e + \dot{\varepsilon}_{ij}^p. \tag{2.23}$$

Garson, proceeding from a theoretical analysis of the problem of the spherically symmetric deformation of a spherical void in an ideal-plastic material, obtained the condition for the plasticity of a porous material in the form

$$f = \left(\frac{\tau}{\sigma_T}\right)^2 + 2q_1 m \operatorname{ch}\left(-\frac{3}{2}\frac{q_2 p}{\sigma_T}\right) - \left(1 - (q_1 m)^2\right) = 0, \tag{2.24}$$

where $\tau = \sqrt{\frac{3}{2} s_{ij} s_{ij}}$ – is the intensity of tangential stresses; $s_{ij} = \sigma_{ij} - \sigma_0 \delta_{ij}$ is the deviator of the Cauchy stress tensor; $\sigma_T(\varepsilon^p)$ is the yield point of a continuous material (matrix material), depending on the intensity of plastic deformation; m is porosity (volume fraction of voids in the material).

Tvergaard [136] introduced the fitting coefficients q_1 and q_2 into the condition (2.24) and, for stretching, using the experiment, obtained $q_1 = 1.5$ and $q_2 = 1.0$.

The plasticity condition (2.24) is taken as the plastic potential, and, in accordance with the associated law, we obtain defining relations of the form

$$\dot{\varepsilon}_{ij} = \lambda \frac{\partial f}{\partial \sigma_{ij}} = \lambda \left(-\frac{1}{3}\frac{\partial f}{\partial p}\delta_{ij} + \frac{3}{2\tau}\frac{\partial f}{\partial \tau} s_{ij}\right), \tag{2.25}$$

where λ is a nonnegative scalar factor.

Strengthening–softening of the matrix material is described by the dependence $\sigma(\varepsilon^p)$. Since the work of plastic deformations is performed only by the matrix material, an evolution equation $\dot{\varepsilon}^p$ is obtained in the form of a power equation:

$$(1-m)\sigma_T \dot{\varepsilon}^p = \sigma_{ij}\dot{\varepsilon}_{ij}^p. \tag{2.26}$$

The change in the porosity of the material is due to the growth of existing voids and the generation of new ones: $\dot{m} = \dot{m}_g + \dot{m}_\tau$.

From the continuity equation, assuming the matrix material is incompressible, we obtain an equation for the growth rate of the voids

$$\dot{m}_g = (1-m)\dot{\varepsilon}^p \delta_{ij}. \tag{2.27}$$

Fundamentals of the Phenomenological Theory of Fracture

It is assumed that the generation of voids occurs when the grains are displaced relative to each other (as a result of grain-boundary slip) and the rate of this process is proportional to the rate of plastic deformation:

$$\dot{m}_\tau = A\dot{\varepsilon}^p. \qquad (2.28)$$

The strain rates at which the voids originate obey the normal distribution law with the mean $\dot{\varepsilon}_N$ and the variance $\dot{s}_N$, i.e.

$$A(\dot{\varepsilon}^p) = \frac{m_N}{s_N\sqrt{2\pi}}\exp\left[-\frac{1}{2}\left(\frac{\dot{\varepsilon}^p - \dot{\varepsilon}_N}{\dot{s}_N}\right)\right], \qquad (2.29)$$

where the maximum volume fraction of nascent voids in the material is m_N. Voids are born only under tensile loading, i.e. $\varepsilon_{ii}^p > 0$.

This model was used to solve practical problems. It agrees well with the experimental results for slow quasistatic loading, correctly describes the plastic compressibility and the dilatancy effect. However, it is not without flaws. The model does not take into account the effect of the strain rate on the damage (the porosity value). The equations of the model do not give a correct description of the damage in the case of materials with falling deformation diagrams.

The author [80, 103] proposed in the mesomechanics a micromechanical multiscale model of the damage of an elastoplastic material. Following [103], we consider the sequence of arguments of the author[1].

The initial stage of deformation is determined by the motion of microdefects of the crystal lattice – dislocations. The rate of viscoplastic deformation is

$$\dot{\varepsilon}^p = \bar{m}b\rho_g v, \qquad (2.30)$$

where $\bar{m}$ is the orientation coefficient; b is the modulus of the Burgers vector of dislocations; ρ_g is the scalar density of mobile dislocations; v is the average slip velocity of dislocations, defined as

---

1) The model considered below is constructed taking into account the physical concepts of plastic deformation and the fracture of metals. Therefore, concepts and representations from the field of strength and plasticity physics are used. In this regard, this material is recommended to be read after studying Chap. 3.

$$v = v_0 \exp\left(-\frac{U_0 - (s - s^r)}{kT}\right), \qquad s \geq s^r. \tag{2.31}$$

Here U_0 is the activation energy; k is the Boltzmann constant; T is the absolute temperature; s^r is residual stress; $(s - s^r)$ is the active stress.

The density of mobile dislocations ρ_g increases in proportion to the plastic deformation ε^p and decreases with increasing total number of dislocations ρ due to their stopping at grain boundaries:

$$\rho_g = (\rho_0 + \alpha \varepsilon^p)^n \exp(-\rho/\rho^*), \tag{2.32}$$

where ρ_0, ρ^* and α are material constants.

To describe the process of nucleation and development of microdefects, the balance of dislocation flows is considered. Denoting: $\dot{\rho}_{ij}$ – the total flow of dislocations at the initial stage of plastic deformation; $\dot{\rho}_{ij}^g$ – the flow of mobile dislocations contributing to deformation; $\dot{\rho}_{ij}^s$ – the flow of dislocations that accumulate on obstacles; η is the part of the flow $\dot{\rho}_{ij}$ associated with mobile dislocations; $(1 - \eta)$ – part of the flow associated with the stationary dislocations, we writte

$$\dot{\rho}_{ij}^g = \eta \dot{\rho}_{ij}, \quad (1-\eta)\dot{\rho}_{ij} = \dot{\rho}_{ij}^s, \quad 0 < \eta < 1.0. \tag{2.33}$$

This implies

$$\dot{\rho}_{ij}^s = (1 - \eta)\dot{\rho}_{ij}^g / \eta.$$

With further development of the deformation, a partial annihilation of the flow of stationary dislocations occurs, disclinations are formed, and grain-boundary slip occurs. As a result of these processes, damage occurs in the form of microcracks and microvoids. At this stage, the balance of flows will be written as

$$(1 - \eta)\dot{\rho}_{ij} = \dot{\rho}_{ij}^s \dot{\rho}_{ij}^0,$$

where $\dot{\rho}_{ij}^0$ is the tensor of the flow of annihilating dislocations.

It is assumed that

$$\dot{\rho}_{ij}^0 = \dot{\lambda}\dot{\rho}_{ij}^s, \qquad (2.34)$$

where $\dot{\lambda}$ is the unknown scalar factor.

To determine the coefficient $\dot{\lambda}$, it is assumed that the annihilation process begins when the intensity of the tensor $\dot{\rho}_{ij}^s$ $\left(\Omega = (3/2\rho_{ij}^s\rho_{ij}^s)^{1/2}\right)$ reaches the critical value Ω_*. In this case, the annihilation intensity $\dot{R}_{ll}$ will be a monotonie function of excess intensity $(\Omega_{ll} - \Omega_*)$, that is,.

$$\dot{R}_{ll} = \left(3/2\dot{\rho}_{ij}^a\dot{\rho}_{ij}^a\right)^{1/2} = \hat{Q}(\Omega_{ll} - \Omega_*)/\tau_p,$$

$$\hat{Q}(\xi) = \begin{cases} Q(\xi), & \xi \geq 0 \\ 0, & \xi < 0 \end{cases}, \qquad (2.35)$$

where $Q(\xi)$ is the dimensionless function of its argument, and τ_p is the dimension parameter [C] associated with the void scale. It follows from (2.35) that

$$\dot{\lambda} = \frac{\hat{Q}(\Omega_{ll} - \Omega_*)}{\tau_p \Omega_{ll}}. \qquad (2.36)$$

The final equation for determining the flow $\dot{\rho}_{ij}^g$ from (2.33)–(2.36) can be written as follows:

$$\frac{d\rho_{ij}^s}{dt} + \frac{\hat{Q}(\Omega_{ll} - \Omega_*)}{\tau_p \Omega_{ll}}\rho_{ij}^s = \frac{1-\eta}{\eta}\frac{d\rho_{ij}^g}{dt}. \qquad (2.37)$$

The author of the model, taking into account the known experimental fact that the value of the microstrain is proportional to the density of the dislocation cluster at the grain boundaries, considers that the macrophysical meaning of the tensor ρ_{ij}^s corresponds to the microstress tensor.

The critical density of dislocations Ω_* at which microcracks are generated on obstacles and the corresponding intensity of residual stresses s_*^r can be determined at the microlevel on the basis of dislocation models of nucleation of microcracks.

The condition $\Omega_{ll} = \Omega_*$ in the stress space has the meaning of the surface of the onset of failure.

The characteristics of the annihilation tensor p_{ij}^a correlate with the damage tensor of the medium D_{ij}. The deviator of the tensor p_{ij}^a is connected with the deviator of the tensor D_{ij}, which leads to the relaxation of the residual stresses (the second term in (2.37)). The spherical part $\dot{p}_{ii}^a$ correlates with the volume deformation of the damage $\dot{D}_{ii}$ and is related to the growth rate of the porosity, which is determined by the continuity equation.

The author proceeds from the foregoing micromodel to the macromodel [103] as follows. Equations (2.30)–(2.32) are written for the three-dimensional stress–strain state in the generalized form in the form of a relation between the second invariants of the plastic strain tensor $\dot{\varepsilon}^p$ and the active stress tensor s^0:

$$\dot{\varepsilon}^p = m(\varepsilon^p) \frac{\psi(s^a - s_0^a)}{\tau_d},$$

$$s^a = \left(\frac{3}{2} s_{ij}^a s_{ij}^a\right)^{1/2}, \dot{\varepsilon}^p = \left(\frac{2}{3} \varepsilon_{ij}^p \varepsilon_{ij}^p\right), \quad (2.38)$$

$$s_{ij} = \sigma_{ij} - \frac{1}{3} \delta_{ij} \sigma_{kk}, \varepsilon_{ij}^p = \varepsilon_{ij} - \frac{1}{3} \delta_{ij} \varepsilon_{kk},$$

where τ is the relaxation time of the active stresses, related to the scale of the dislocations.

From equations (2.20)–(2.32) for the first stage of deformation, when $\Omega_{ll} < \Omega_*$, on the basis of the hypotheses of the flow theory, we obtain elastoviscoplastic equations for the volume deformed state:

$$\dot{\varepsilon}_{ij} = \frac{1}{2\beta} \dot{s}_{ij} + m(\varepsilon^p) \frac{\psi(s^a - s_0^a)}{\tau_d T} s_{ij}^a,$$

$$\sigma_{ii} = 3K\varepsilon_{ii}, \varepsilon_{ij} = \varepsilon_{ij} - \frac{1}{3} \delta_{ij} \varepsilon_{kk}. \quad (2.39)$$

Assuming, for simplicity, the material is plastically incompressible: $\varepsilon_{kk}^p = 0$, for the spherical parts of the tensors ε_{ij}^p and σ_{ij} we obtain an elastic law.

If we assume that for the deviators of the residual stresses and the accumulated dislocation flow,

$$s_{ij}^r = 2\mu^* \rho_{ij}^{s'}, \quad (2.40)$$

then for the condition $\Omega_{ll} < \Omega_*$ we get that the residual stresses s_{ij}^r

Fundamentals of the Phenomenological Theory of Fracture 75

satisfy the condition $s^r \leq s_0^r$, where μ^* is the constant of the voltage dimension. The evolution equation (2.33) for s_{ij}^r in the first stage of plastic deformation takes the form

$$\frac{1}{2\mu^*}\dot{s}_{ij}^r = \frac{1-\eta}{\eta}\dot{\varepsilon}_{ij}^p. \qquad (2.41)$$

Integration (2.41) with respect to time for $\mu^* = $ const gives the well-known kinematic hardening law

$$s_{ij}^r = 2\alpha\varepsilon_{ij}^p, \alpha = \mu^*\frac{1-\eta}{\eta}. \qquad (2.42)$$

The hardening modulus α is determined from the experimental data on the Bauschinger effect.

Thus, before the formation of microvoids, the matrix material is described in accordance with the theory of dislocations by the equations (2.39)–(2.42) of an elastoviscoplastic medium with kinematic hardening. When microcracks and microvoids $(s^r \geq s_0^r)$ appear for the tensor s_{ij}^r, the following relaxation equation is obtained

$$\dot{s}_{ij}^r + \frac{2\mu^*}{\tau_p}\frac{Q(s^r - s_0^r(m))}{s^r}s_{ij}^r = 2\alpha\dot{\varepsilon}_{ij}^p. \qquad (2.43)$$

Equations (2.41)–(2.43) can be written as a single equation by introducing the function $\hat{Q}(\xi)$ in accordance with formula (2.35) This equation describes the relaxation of residual stresses after the beginning of the process of formation of microvoids, and consequently, softening of the material. in accordance with equation (2.42). It follows from (2.43) that the relaxation occurs up to a certain stationary value $s_0^r(m) \neq 0$. Further, for $t \gg \tau_p$, the residual stresses will vary according to the law of plastic flow associated with the yield surface and depending on the porosity $s^r = s_0^r(m)$ which follows from (2.43) if $\tau_p \to 0$

$$ds_{ij}^r + H\left(s_{mn}^r d\varepsilon_{mn}^p\right)\frac{s_{mn}^r d\varepsilon_{mn}^p}{\left(s_0^r(m)\right)^2}s_{ij}^r = 2\alpha\varepsilon_{ij}^p, \qquad (2.44)$$

where m is the porosity; $H(\xi)$ is the Heaviside function.

With the appearance of damage, the situation changes significantly. The material becomes two-phase – consists of a matrix and voids.

Knowing the properties of the matrix, which are described by the equations (2.39), (2.42), one can find the effective characteristics of a plastic material. However, with the appearance of voids, the plasticity condition of the material changes, and equations describing the nucleation and evolution of defects are needed to describe the deformation. The difficulty of solving this problem depends on the accepted void geometry.

As an illustration, let us consider the solution of the problem of a unit void in a medium under the action of forces applied at infinity, and then summing this solution for some known void size distribution and orientation.

Within the framework of this approach, an analytical solution was obtained only for simple linear media.

For a perfect elastoplastic porous material with voids of simple spherical shape, the Garson plasticity condition is adopted. It is generalized to a porous material, the matrix of which is described by elastic-viscoplastic equations with kinematic hardening

$$\Phi\left(\sigma_{ij}^{a}, m, \sigma_{T}\right) = \frac{3}{2}\frac{s_{ij}^{a}s_{ij}^{a}}{(\sigma_{T})^{2}} + 2q_{1}mch\frac{3q_{2}}{2}\frac{\sigma_{kk}^{a}}{\sigma_{T}} - \left[1 + (q_{1}m)^{2}\right] = 0, \quad (2.45)$$

where s_{ij}^{a} is the deviator of the tensor of active stresses, σ_{T} is the yield stress of the matrix of elastic-viscoplastic material, which is determined from the condition of equality of plastic work for the matrix and effective material:

$$\sigma_{ij}^{a}\varepsilon_{ij}^{p} = (1-m)\dot{\varepsilon}^{p}\left[\sigma_{T}(\varepsilon^{p}) + \psi^{-1}(\tau\dot{\varepsilon}^{p})\right],$$

$$\sigma_{T} = \sigma_{T}(\varepsilon^{p}) + \psi^{-1}(\tau\dot{\varepsilon}^{p}). \quad (2.46)$$

The first equation is used to determine ε^{p}, after which from the second equation σ_{T}. The stress σ_{ij}^{0} and the rate of dissipative strains $\dot{\varepsilon}_{ij}^{p}$ in the effective material are related by the associated law of plastic flow

$$\dot{\varepsilon}_{ij}^{p} = \dot{\lambda}\frac{\partial F}{\partial \tau_{ij}}, \quad (2.47)$$

if the associated law holds for the material of the matrix. The parameter $\dot{\lambda}$ is determined from the second equation (2.46). From the continuity equation follows the equation of growth of voids

Fundamentals of the Phenomenological Theory of Fracture 77

$$\dot{m}_{gr} = (1-m)\dot{\varepsilon}_{kk}^p = \Lambda \frac{3m(1-m)}{\sigma_T} q_1 q_2 sh \frac{3q_2 \sigma_{kk}}{2\sigma_T}. \tag{2.48}$$

The equations of nucleation and evolution of voids remain the same as in the GTN model (2.26)–(2.29). They complete the system of defining equations (2.45)–(2.48). From them, the stresses and internal parameters of the material are determined using the velocity field found from the conservation laws.

The criterion of fracture is the equality of the void density to the critical value. When the porosity reaches a critical value $m = m_*$, a catastrophic void spreading and material fracture occurs.

The value of m_* depends on many external factors: temperature, loading rate and others, and also on the structure of the material and, as experiments show, varies within 0.05–0.50.

The stated connected model of plasticity and damage, in contrast to the GTN model, is multiscale. It takes into account three temporal and, respectively, three spatial scales: t_0 is the characteristic time of the problem, τ_p is the relaxation time of the stress in the damaged material, τ_d is the relaxation time in the initial viscoplastic material, corresponding respectively to the macro-, meso- and microscales. Wherein $t_0 \gg \tau_p \gg \tau_d$.

At the hardening stage, the scale effect is determined by the small parameter $\delta_d = \tau_d/t_0$, and at the softening stage $\delta_p = \tau_p/t_0 \gg \delta_d$. As δ_p and δ_d tend to zero, the model goes over into the GTN model with kinematic hardening that does not depend on the scale factors.

Analysis of the model described allows us to make the following conclusion.

1. The micromechanical connected multiscale model of plastic deformation and ductile fracture proposed by V.N. Kukudzhanov is one of the first connected models consistently constructed with the involvement of the mechanics and physics of the strength and plasticity of metals, i.e., it is a synthetic model.

2. In the model, the damage is described by a tensor quantity, it takes into account kinematic hardening and describes the Bauschinger effect.

3. In the construction of the model, some assumptions that have been accepted contradict the available experimental data. So, for example, it is assumed that at the first stages of deformation the voids are still absent and the material behaves as a hardening medium. Beginning with some deformation, voids appear in the material and, consequently, processes of localization and softening develop. It has been experimentally proved that the generation of

submicrocracks, that is, the appearance of damage, in metals occurs almost simultaneously with the appearance of residual deformations [72].

It is also assumed that the density of mobile dislocations ρ_g increases in proportion to the degree of plastic deformation ε_p and the equation (2.32) is written. In this case, equation (2.30) was previously adopted according to which the density ρ_g is proportional not to ε_p, but to $\dot{\varepsilon}_p$. It is known that the value of the accumulated plastic deformation is proportional to the density of dislocations transmitted through the deformable volume [70].

4. The analyzed model, like the GTN model, does not take into account the 'healing' of voids, i.e., the reduction of damage in soft stress states.

5. The foregoing shortcomings, apparently, are conditioned by the fact that the model only qualitatively allows to predict the deformation and accumulation diagrams of metal damage [103].

Thus, despite the rather intensive development of the theory of ductile fracture of metals at large plastic strains, the problem of accounting for the history of loading, as in the case of the theory of irreversible strains, remains unresolved.

Plastic deformation and fracture, as has become obvious, is a single physical process. Therefore, the approach to its description should be the same. *Therefore, a complete description of the irreversible deformation must include, within the framework of a single model, the deformation law (determining equations) taking into account the loading history, and the law of accumulation of deformation damage with the criterion of ductile fracture, which also takes into account the load history.* At the same time, the mathematical formulation of these laws should permit the solution of practical problems by numerical methods.

The promising direction of the solution of the formulated scientific problem, as follows from the analysis of published works, is the creation of synthetic models based on the basic principles of mechanics and the physics of strength and plasticity of metals.

In the following chapters one of the possible variants of such a model is described. To ensure the clarity and consistency of the exposition, let us briefly dwell on the basic concepts, fundamental principles and method in the physics of strength and plasticity of metals.

3

Fundamentals of the physics of strength and plasticity of metals

3.1. Basic concepts and assumptions of the dislocation theory of plasticity

The microstructural description of plastic deformation and fracture is based on a *discrete (atomic) model* of metallic materials. Under normal production conditions, metals and metal alloys have a *crystalline structure*, the characteristic feature of which is the presence of a long-range order. The concepts of an *ideal crystal lattice* and its *defects* are introduced, which are subdivided into defects of the atomic level and *microdefects*. The first include *phonons* – quanta of elastic waves in a crystal and *point defects*, whose dimensions in three mutually perpendicular directions are of the order of interatomic distances. These are *vacancies, interstitial atoms, interstitial and substitutional atoms*, their simplest complexes and combinations.

Microdefects are defects the dimensions of which in one or two dimensions are close to the interatomic dimensions, and in the other two (or in the other) are much larger. Accordingly, *planar and linear* microdefects are distinguished. The first include *stacking faults*, grain boundaries and interphse boundaries. Of particular importance for the physics of strength and plasticity are linear defects – *partial and complete dislocations*.

Phonons are the collective thermal motion of atoms in a crystal in the form of elastic waves [67]. The speed of their propagation is equal to the speed of sound c in the material. The crystal lattice has a wide set of frequencies v and, respectively, wavelengths $\lambda = c/v$.

The minimum wavelength in the lattice is $\lambda_{min} = 2b$, where b is the interatomic distance. The limiting frequency of the thermal vibrations of the ion at the lattice site – the Debye frequency $v_D = c/2b$ – is practically independent of the temperature and for metals lies in the interval $v_D = 10^{12}-10^{13}$ s^{-1} [67]. The energy of thermal oscillations with frequency v is hv, where h is the Planck constant.

Thermal vibrations of the lattice play a very important role in the formation of the properties of metals. Fluctuations in the energy of these oscillations determine the thermal fluctuation mechanisms of the formation and motion of point defects, the movement of dislocations, the formation and healing of microcracks during plastic deformation.

Vacancy – the main point defect – is a vacant lattice node. An *interstitial atom* is an atom located in the interstices of a lattice. The concentration (density) of vacancies in a certain volume V of a crystal is determined by the ratio of the number of vacancies n to the total number of nodes in the volume

$$c_v = n/(N+n), \tag{3.1}$$

where N is the number of sites filled with atoms.

In the local volume of a crystal with a vacancy, as with an interstitial atom, the interatomic distances are changed and differ from the equilibrium distances. This causes the presence of a field of displacements and, accordingly, stresses equilibrated in a certain volume, which also determine the presence of excess elastic energy. This energy rapidly decreases with the distance from the defect. The vacancy energy in *lattice units* Gb^3 is equal to $U_v = \beta_v Gb^3$, where β_v for different metals lies in the range of 0.14–0.33. The estimates take $\beta_v = 0.25$, that is, $U_v \approx Gb^3/4$ [67].

Vacancies and interstitial atoms are *equilibrium lattice* defects. Therefore, the dependence of their concentration in a certain volume of the crystal on the temperature at normal pressure, with allowance for (3.1), is described by the Boltzmann equation [67]:

$$c_v(T) \approx \frac{N}{N}\exp\left(-\frac{U_v}{kT}\right) \approx \exp\left(-\frac{U_v}{kT}\right), \tag{3.2}$$

where $\exp(-U_v/kT)$ is the probability of the thermal fluctuation formation of a vacancy (the escape of an atom from a site); N in the numerator is the number of probable places of formation of vacancies, equal to the number of all atoms in the volume, performing thermal vibrations; U_v is the activation energy of the formation of a vacancy, equal to the vacancy energy $U_v = Gb^3/4$; k is the Boltzmann constant.

The *activation volume* (the volume of the lattice that needs to be activated to form a vacancy) is approximately equal to the volume of vacancy b^3. The positive hydrostatic pressure p increases the energy of formation of the vacancy and (3.2) in this case has the form

$$c_v(T) \cong \exp\left(-\frac{U_v + pb^3}{kT}\right). \qquad (3.3)$$

If, due to thermal fluctuations, the amplitude of the oscillations of an atom located near a vacancy exceeds $b/2$, it passes through a potential barrier of height U_m, descends to the neighbouring potential well $x = b$ and takes the place of vacancy, i.e., the atom and vacancy are exchanged (Fig. 3.1) [67]. This process is an elementary act of *vacancy migration* and at the same time an elementary act of *self-diffusion* of matter. U_m is the activation energy of vacancy migration. Adjacent to the vacancy the atom performs thermal oscillations with a frequency $v \approx v_D$, and the energy U_m can acquire with probability $\exp(-U_m/kT)$. Consequently, the average frequency of the elementary process of vacancy migration is

$$v_v = v_D \exp(-U_i/kT), \qquad (3.4)$$

where the Debye frequency v_D is the meaning of the number of attempts of the atom per unit time to jump through the potential barrier, and $\exp(-U_m/kT)$ is the probability of overcoming the barrier.

To implement the elementary act of self-diffusion, the formation of a vacancy is necessary, and, consequently, the energy cost U_v,

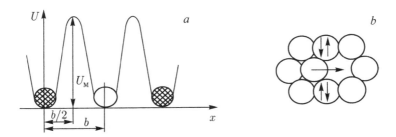

Fig. 3.1. Elementary act of vacancy migration: a – potential barrier; b – displacements of adjacent atoms.

and then the neighbouring atom must take its place, breaking the energy barrier U_m. Therefore, the activation energy of self-diffusion is $U_c = U_v + U_m$. For most metals, $\beta_m/\beta_v = 0.5-0.7$ [67]. Then $\beta_m \approx 0.25 \cdot 0.6 = 0.15$ and $U_c = Gb^3/4 + 3Gb^3/20 = Gb^3\left(\frac{1}{4}+\frac{3}{20}\right) \cong Gb^3/2.5$. This is the estimate of the average value.

Equations of the type (3.2)–(3.4) describe elementary processes in crystalline bodies for which the activation energy is of the same order as the energy of thermal vibrations of atoms kT per one degree of freedom. If $U \gg kT$, then the migration process is impossible, since its probability is very small. For this reason, the described processes are practically not observed at very low temperatures, when kT is small.

The point lattice defects and the associated thermally activated processes play an important role in a single process of plastic deformation and destruction of metals, in particular, in the case of *non-conservative dislocation movement*, the formation and healing of microcracks.

Experience shows that the crystal lattice is a very strong construction. It can be completely destroyed only by melting. In the case of plastic deformation of mono- and polycrystals, the irreversible displacement of some lattice volumes relative to others with its conservation occurs due to a special mechanism of motion of linear defects – *dislocations*. The physics of strength and plasticity as an independent scientific discipline originates from the introduction of the concept of dislocations in 1934 and the identification of their properties.

The theory of dislocations is described in detail in many textbooks and monographs [67–70, 138–141]. We will briefly dwell on the concepts, effects and positions that are necessary for the presentation of further material.

The successive stages of sliding of the edge branches of a dislocation in the slip plane A–A under the action of the stress τ are shown in Fig. 3.2 *d–f*. While the shear has not reached the crystal boundaries, at every moment it is possible to reveal the boundary of the shear region in the slip plane – the dislocation (line $ACBD$) (Fig. 3.2 *a*, *b*). Consequently, the *dislocation is the perimeter of the plastic shear area in the slip plane*.

The dislocation line (dislocation axis) is designated in the diagrams as follows: ⊥ – positive dislocation and ⊤ – negative dislocation. It is always closed (forms a loop), or its edges emerge on the free surface of the crystal. The main characteristics of

Fundamentals of the Physics of Strength and Plasticity

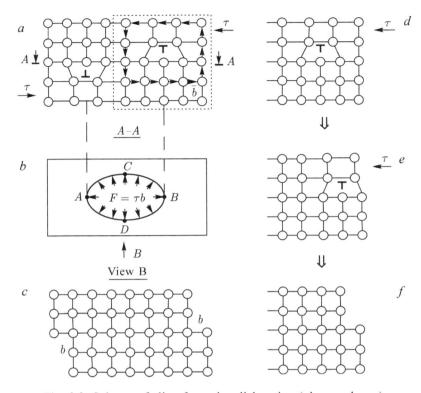

Fig. 3.2. Scheme of slip of an edge dislocation (planar scheme).

dislocations are: the *Burgers vector b*, the *scalar density* ρ, the *average dislocation distance* $l = 1/\sqrt{\rho}$, *energy of a unit of length* (in estimates is taken equal to $Gb^2/2$).

The *Burgers vector* closes the mental contour, which is formed when the dislocation axis is traversed, as shown in Fig. 3.2 *a*. For an edge dislocation, it describes the direction of its slip and power. Its modulus is almost equal to the lattice parameter, and for metals, $b \approx (2-4) \cdot 10^{-8}$ cm.

The *scalar dislocation density* is the total length of the dislocation lines per unit volume of the crystal – $\rho \left[\dfrac{\text{cm}}{\text{cm}^3} \right] = \left[\text{cm}^{-2} \right]$. In annealed metals, $\rho \approx 10^8$ cm^{-2}, in the highly deformed metals the maximum density is $(10^{12}-10^{13})$ cm^{-2}. Dislocations are a non-equilibrium lattice defect, so annealing does not destroy them completely. They arise in metals during crystallization of the melt and during plastic deformation.

In the presence of a stress τ in the lattice plane (in the sliding plane of the dislocation), the force $F = \tau b$ (Fig. 3.2 a, b) acts on its unit length and it directed at each point along the normal to the dislocation line in the direction to which the shear has not yet passed. Dislocations interact with each other and with other lattice defects by means of their long-range elastic fields. The result of the interaction is determined by the law of the system's striving for an equilibrium state with a minimum of free energy. In a non-stressed crystal, dislocations of the opposite sign are spontaneously attracted and *annihilated* (destroyed). The dislocation loop contracts, shrinks to the point and self-destructs. Its elastic energy dissipates, transforming into energy of thermal quanta – phonons.

When a dislocation runs through the entire slip plane, with an exit to the crystal boundary, an elementary shift occurs – an irreversible displacement of one part of the crystal relative to the other by one interatomic distance (Fig. 3.2 c). This is an elementary act of plastic deformation. The macroscopic shift is the result of the run of many dislocations. The successive rupture with the subsequent restoration of interatomic bonds during the slip of the edge dislocation (Figs. 3.2 d–f) causes a smaller (by one or two orders of magnitude) value of the real strength of metals in comparison with the theoretical shear strength $\tau_{max} \approx G/2\pi$, with simultaneous rupture of all interatomic bonds in the slip plane.

When the dislocation slides, its energy changes with a period equal to the lattice parameter. In the potential well (Fig, 3.2 d), the dislocation energy is minimal. The transition to the neighbouring potential well (Fig. 3.2 e) is connected with overcoming the potential barrier at the moment of discontinuity and restoration of interatomic bonds. These periodic barriers are called *Peierl's relief*. The set of the crystallographic plane and the dislocation slip direction constitute a sliding system. Peierl's relief is minimal in the crystallographic planes and directions, most densely packed with atoms. These systems are called *easy slip systems*.

Figure 3.3 shows easy slip systems in the main types of crystal lattices (the sliding planes are shaded, and the arrows indicate the sliding directions).

Each sliding system is characterized by its critical stress τ_{cr}, at which sliding begins. This stress does not depend on the orientation of the slip system with respect to the acting stress, which determines the dependence on the orientation of the yield stress of a single crystal, i.e., the *anisotropy of mechanical properties*.

Fundamentals of the Physics of Strength and Plasticity   85

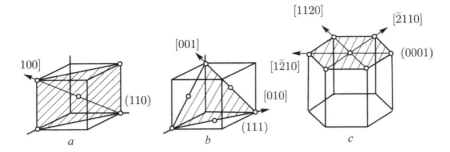

Fig. 3.3. Some systems of easy slip in lattices: *a* – cubic body-centered (BCC); *b* – cubic face-centered (FCC); *в* – hexagonal close-packed (HCP) (shown on one plane of easy slip plane and in it several directions of sliding).

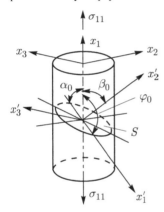

Fig. 3.4. Derivation of the Schmid law.

Let us consider a single crystal in a uniaxial stress state. The tensile normal stress σ_{11} acting in the direction of the x_1 axis is the only component of the stress tensor σ_{ij} in the coordinate system x_1, x_2, x_3 (Fig. 3.4).

We choose a new coordinate system x'_1, x'_2, x'_3 so that the normal to the slip plane S coincides with the x'_2 axis. We denote the angle between the x'_2 axis and the direction σ_{11} (x_1 axis) as β_0, and the direction cosine between the axes x'_2 and x_1 through $\lambda_{2'1} = \cos\beta_0$. The slip direction in the new coordinate system coincides with the direction of the x'_1 axis and makes an angle α_0 with the direction σ_{11} (the direction cosine $\lambda_{1'1} = \cos\alpha_0$). The stresses in the considered sliding system in the new coordinates x'_1, x'_2, x'_3 are obtained by transforming the tensor σ_{ij} by the formula $\sigma_{i'j'} = \lambda_{i'i}\lambda_{j'j}\sigma_{ij}$. Dislocation slip is inherently a shear mechanism of deformation and occurs under the action of tangential stresses. The tangent component of the tensor $\sigma_{i'j'}$

$$\sigma_{1'2'} = \tau = \lambda_{1'i}\lambda_{2'j}\sigma_{ij} = \lambda_{1'1}\lambda_{2'1}\sigma_{11} = \cos\alpha_0 \cos\beta_0 \sigma_{11}. \qquad (3.5)$$

The multiplier $m' = \cos\alpha_0 \cos\beta_0$ is called the *Schmid factor*. It follows from (3.5) that if the direction of σ_{11} is parallel to the slip plane S ($\beta_0 = 90°$) or normal to it ($\beta_0 = 0$ and $\alpha_0 = 90°$), the stress $\tau = 0$ and sliding in this plane can not occur. The maximum tangential stress $\tau_{max} = 0.5\sigma_{11}$ in the slip plane occurs when $m' = 0.5$ and, accordingly, $\alpha_0 = \beta_0 = 45°$. If the slip in the plane has begun, therefore, $\tau = \tau_{cr}$ and $\sigma_{11} = \sigma_T$. From (3.5) it follows that

$$\sigma_T = \tau_{cr}/m' = m\tau_{cr}, \qquad (3.6)$$

where $m = 1/m'$ is the reciprocal of the value of the Schmid factor. For each sliding system, $\tau_{cr} = $ const. Consequently, σ_T depends on m, that is, on the orientation of the slip systems with respect to the normal stress σ (Schmid law).

A significant mechanism for plastic deformation and fracture microprocesses is the mechanism of overcoming the dislocation of the Peierls barriers in the slip planes by means of pairwise inflections (Fig. 3.5) [70]. It consists of two stages. On the first, a pairwise inflection occurs (Fig. 3.5 b). On the second – its distribution in both directions (Figs. 3.5, c, d).

The initiation of a paired inflection is a thermal fluctuation process. Therefore, the probability of its formation is proportional to the Boltzmann factor

$$P \sim \exp(-U/kT), \qquad (3.7)$$

where U is the inflection energy. In the case of plastic deformation, the potential barrier for the formation of a pairwise inflection decreases due to the work of stresses, and the probability of its formation increases:

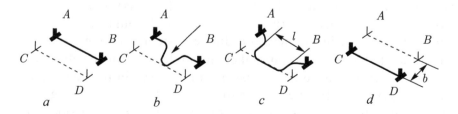

Fig. 3.5. Dislocation slip by the mechanism of pairwise inflection: *a* – the starting position; *b* – nucleation of a pairwise inflection; *c* – the spread of a pair inflection; *g* – final position – dislocation displacement by one interatomic distance *b*.

Fundamentals of the Physics of Strength and Plasticity   87

$$P \sim \exp[-(U - \tau V)/kT] = \exp[-(U - \tau b^2 l)/kT], \qquad (3.8)$$

where $V = b^2 l$ is the activation volume, l is the length of the inflection.

The expansion of the inflection into one interatomic distance occurs over a time $t = v_D^{-1}$ with a velocity equal to the speed of sound c in the metal.

In addition to sliding (conservative motion), an edge dislocation can perform a non-conservative motion – *climb*, as shown in Fig. 3.6. When climbing, the dislocation passes into equivalent parallel slip planes with the emission of interstitial atoms or vacancies. With this mechanism, an edge dislocation can bypass barriers and continue sliding in another equivalent plane.

Climb is a thermally activated process controlled by diffusion. Therefore, its activation energy is equal to the activation energy of self-diffusion. Bypassing the edge dislocations of barriers by climb and converting them from stationary to mobile or annihilation with a dislocation of the opposite sign is one of the mechanisms of *thermal recovery*.

If the dislocation is at the points A and B (Fig. 3.2 b), then at the points C and D it is screw, and in the intervals – mixed. A screw dislocation has a number of specific properties (Fig. 3.7).

If the edge dislocation **b** is perpendicular to its line, then the screw disposition is parallel. The edge dislocation has one slip plane, there are many of them. The dislocation line AB (Figure 3.7) can change the direction of sliding and, accordingly, the gliding plane. For this reason, the screw dislocation is more mobile from the point of view of easy bypassing of barriers by the transition of its section from one slip plane to another – by the cross slip mechanism (Fig. 3.8) [70]. The cross slip of screw dislocations as a mechanism for traversing of barriers by the dislocations plays an important role in plastic deformation. It ensures the transformation of stationary

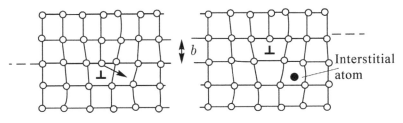

Fig. 3.6. Climb of an edge dislocation with the emission of interstitial atoms (a planar scheme).

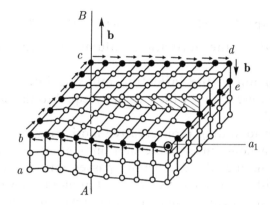

Fig. 3.7. Screw dislocation in the crystal lattice: AB – dislocation line; **b** – Burgers vector.

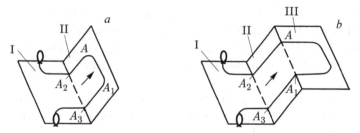

Fig. 3.8. Cross slip (*a*) and double cross slip (*b*) of a screw dislocation: I – the initial slip plane; II – cross slip plane; III – slip plane parallel to the original.

dislocations (sitting on barriers) into mobile dislocations, that is, it is a *recovery mechanism*.

During plastic deformation, dislocations multiply (their density ρ increases). The main multiplication mechanism is the Frank–Read mechanism known in the literature as the Frank–Read source. The segment of the dislocation line AA_1 (Fig. 3.9), most often a half loop of a screw dislocation that has undergone double cross slip (Fig. 3.8 *b*), under the action of force F bends with increasing length and emits a series of dislocation loops in the slip plane. The stress of the source is directly proportional to the Burgers vector modulus of the dislocation segment AA_1 and inversely proportional to its length L (Fig. 3.9).

Earlier it was noted that the easy slip systems are sets of planes and directions that are most densely packed with atoms (Fig. 3.3), and therefore the Burgers vectors of dislocations in them are minimal. The crystal structures of most industrial metals and alloys (BCC, FCC, HCP) have a different number of such systems

Fundamentals of the Physics of Strength and Plasticity

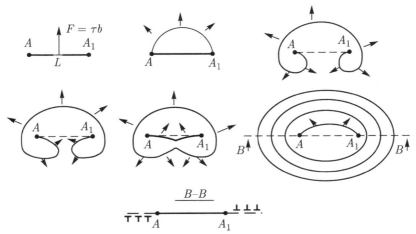

Fig. 3.9. Sequential stages of operation of the Frank–Read source in the slip plane.

and their different orientations. In BCC and FCC crystals, these systems, as is well known [70, 139], intersect, and consequently the dislocations sliding in them also intersect. The act of intersection of the dislocations plays an important role in the process of plastic deformation and causes differences in the plastic behaviour of the above crystals.

Figure 3.10 shows the initial (before the intersection) position (a, c) and the result of the intersection of (b, d) of the edge dislocations [70]. In intersections of dislocations with mutually perpendicular Burgers vectors $\mathbf{b}_1$ and $\mathbf{b}_2$ a step with the Burgers vector $\mathbf{b}_s$ is formed on a stationary dislocation 2 (Fig. 3.10 b). Due to the invariance (conservation) of the Burgers vector along the dislocation line $\mathbf{b}_s = \mathbf{b}_2$ and $\mathbf{b}_s \| \bar{b}_2$. With this $s's = |b_1|$ and the step has an edge orientation.

If the intersecting dislocations have the vectors $\mathbf{b}_1$ and $\mathbf{b}_2$ coinciding in direction (Fig. 3.10 c), then after intersection, the inflections pp' are formed on each dislocation line (Fig. 3.10 d). In this case, the length of the inflection $p'_1 p_1 = b_p = b_1$, and the inflection $p_2 p'_2 = b_p = b_2$. Both vectors $\mathbf{b}_p$ lie in slip planes of dislocations, that is, the kinks have a screw orientation.

The formation of steps and kinks at the intersections of dislocations increases their length and, consequently, energy. Therefore, the sliding of intersecting dislocations requires additional expenditure of energy and an increase in the applied stress. In addition, during deformation, mobile dislocations stop at insurmountable barriers, or as a result of interaction between themselves (dislocation reactions), occupy the positions under which their Burgers vectors do not lie

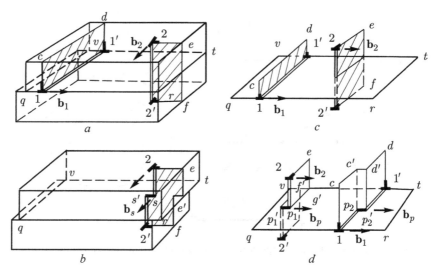

Fig. 3.10. The intersection of orthogonally oriented edge dislocations *1* and *2* with mutually perpendicular (*a*, *b*) and parallel (*c*, *d*) Burgers vectors b_1 and b_2: *qvtr* is the slip plane of dislocation 1; 1*cd*1′, 2*ef*2′ are the extra-planes of dislocations 1 and 2, respectively; *s′s* – step; $p'_1 p_1$ and $p'_2 p_2$ – kinks on dislocations.

in the slip planes. The formed stationary dislocations ('forrest' dislocations) become barriers for mobile dislocations. They repel the moving dislocations of a single sign (long-range interaction) approaching them, by elastic fields, or moving dislocations are forced to cross them (short-range interaction).

Continuation of deformation under these conditions requires a continuous increase in stresses and, consequently, external deforming forces. *The phenomenon of increasing the stresses necessary to continue plastic deformation is called hardening.*

Plastic deformation of materials under uniaxial tension is described by a rheological equation connecting the stress with strain $\sigma(\varepsilon)$ at different temperatures and strain rates. The graphical representation of this dependence in the literature is called differently: *the strain curve, the hardening curve, and the tension diagram*. The tension diagram integrally displays the physical processes occurring in materials under deformation.

Figure 3.11 shows a typical tensile loading diagram of single crystals. In the general case, three stages of deformation are distinguished on it. In stage I, the hardening intensity, characterized by the *hardening modulus* $\theta_1 = \partial\sigma/\partial\varepsilon$, is small and the hardening is practically linear. The deformation begins with the slip of dislocations

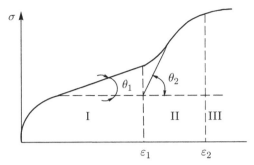

Fig. 3.11. Three-stage strain diagram.

in eaqsyslip systems, their density is still small, and they do not intersect. As the deformation increases, ρ increases, and slight hardening is explained by the long-range interaction of dislocations. This stage is called the *easy slip stage*. With increasing ε and ρ, the stress σ increases and, in accordance with (3.6), new systems come into operation that are least favourably oriented to slip ($m < 0.5$). Movable dislocations begin to intersect, stationary forrest dislocations are formed in parallel, their density increases rapidly with increasing ε, the intensity of hardening sharply increases ($\theta_2 > \theta_1$). Stage II is the *multiple slip stage*. By the beginning of stage III σ, the screw dislocations acquire the ability to bypass the barriers by cross slip (*dynamic recovery*) due to stress work. The intensity of hardening gradually decreases. Stage III is the stage of *dynamic recovery*.

Numerous studies have established [69] that the deformation resistance due to the above-described interaction of dislocations is satisfactorily described by Eq.

$$\tau_{\text{loc}} = \alpha G b \sqrt{\rho}, \tag{3.9}$$

where α is the interdislocation interaction parameter, ρ is the total dislocation density.

Mainly alloys are used as structural materials. Melting and alloying of result in the formation of solid substitution or interstitial solutions. Of the known three types of interaction of dislocations with the atoms of dissolved elements [70] from the point of view of practice, the first type is of special interest, when dislocations are mobile, and the atoms of the dissolved elements are immobile (plastic deformation of solid solutions).

The deformation resistance due to the given type of interaction is defined as

$$\tau_{\lim} = zG\varepsilon_\alpha^{3/2}c^{1/2}, \qquad (3.10)$$

where $z = 1/760$; $\varepsilon_\alpha = R' + \alpha_0 r$ is the parameter of the discrepancy between the sizes of atoms and elastic moduli; $R' = R + |R|/2$; $\alpha_0 = 3$ for an edge dislocation and $\alpha_0 = 16$ for a screw dislocation; c is the concentration of the alloying element; $r = (1/a)da/dc$; a is the lattice parameter; $R = (1/G)dG/dc$.

Comparison of (3.9) and (3.10) implies the identical nature of the dependence of τ on ρ and the concentration of the alloying element c, namely $\tau \sim \rho^{1/2}$ and $\tau \sim c^{1/2}$. The stress $\tau_{\lim}$ does not depend on ε, therefore contributes to σ_T and does not affect the strain hardening intensity.

In dispersion-hardened alloys, the deformation resistance due to dispersed particles of the second phase is associated with the direct interaction of mobile dislocations with particles that are barriers for them and is described by the Orowan equation [70]

$$\tau_{d.p.} \approx \frac{0.6Gb^3\sqrt[3]{f_v}}{r}, \qquad (3.11)$$

where f_v is the volume concentration of the strengthening phase; r is the radius of the particles.

It follows from (3.11) that the more dispersed the particles (less than r) and the greater their volume concentration, the higher $\tau_{d.p.}$.

The resistance of deformation from the interaction of mobile dislocations with point defects, as in the case (3.10), is proportional to the square root of their concentration, i.e., $\tau_{p.d.} \sim c^{1/2}$.

Stresses $\tau_{p.d.}$ and $\tau_{d.p.}$ contribute to the yield strength and do not affect the nature of strain hardening.

The described interactions are thermally activated and are considered additive. Then the macroscopic stress of the monocrystal flow can be written as the sum:

$$\tau_o = \tau_{P-N} + \tau_{loc} + \tau_{\lim} + \tau_{d.p.} + \tau_{p.d.}. \qquad (3.12)$$

From this we can conclude that the dislocation density is mainly influenced by the strain hardening behaviour.

Polycrystals, such as industrial alloys, differ from monocrystals by the presence of grain boundaries and different grain orientation as small, irregularly shaped single crystals. Each grain, like a single crystal, has anisotropic properties. In the case of random orientation of grains in a polycrystal, the latter has isotropic properties in the macroscale. The grain boundaries are practically insurmountable barriers for dislocations. It is established that the flow stress for polycrystals is described by the Hall–Petch relation

$$\sigma_T = \bar{m}\tau_0 + k_y d^{-\frac{1}{2}} = \sigma_0 + k_y d^{-\frac{1}{2}}, \tag{3.13}$$

where $\bar{m}$ is the analog of the Schmid factor m for a single crystal (3.6), called the Taylor factor, averaged for a polycrystal with chaotic grain misorientation; k_y is the constant associated with the propagation of deformation through grain boundaries; d is the average linear grain size. For HCP polycrystals with chaotic misorientation of grains $\bar{m} \cong 2.8 - 2.9$, for polycrystals of fcc $\bar{m} \approx (3.06 - 3.1)$.

As a result of a large directional plastic deformation, especially cold, the grains acquire a predominant orientation and the polycrystal becomes anisotropic. In this case, $\bar{m}$ depends on the direction in the polycrystal. The first term in (3.13) is the friction of the lattice, the second is due to the strengthening effect from the grain boundaries. The latest studies have shown that d, in general, should be understood as the mean free path of dislocations λ (from the source to the barrier).

The presence in metals of barriers for mobile dislocations, as already noted, causes the phenomenon of strain hardening. Barriers are subdivided into *large-scale* ones (hardening is a consequence of long-range internal stress fields), *medium-scale* (consequence of interaction of dislocations in parallel systems), *small-scale* (interactions with point defects and fine particles of second phases, intersections of dislocations, compression of split screw dislocations). On the barriers, dislocations stop and become stationary.

The barriers of the first two types are high ($U \gg 1$ eV) and are overcome due to the acting stresses [72]. Seated on the barriers of the third type, the stationary dislocations can overcome them due to thermal activation and stress (by mechanisms of climb, double cross slip, intersection of forest dislocations) and again become mobile.

The process reverse to the hardening process, associated with a decrease in the number of barriers and their height, which leads to a reduction in the flow stress, is called *softening* or *recovery*. When there are no applied stresses, the recovery is due to thermally activated processes and is called *static* or *thermal* (for example, *static polygonization* and *recrystallization*). The recovery occurring during plastic deformation, when the height of the barriers is lowered by the work of stresses, is called *dynamic*.

The frequency of overcoming small-scale barriers by the dislocation during deformation is described by a modified Boltzmann equation (3.4)

$$v = v_0 \exp\left(-\frac{U - \tau V}{kT}\right), \tag{3.14}$$

where $v_0 \approx$ const is the frequency factor; τV is the stress work; τ is the intensity of tangential stresses; V is the activation volume.

Equation (3.4) describes an equilibrium process, and equation (3.14) is a non-equilibrium (dynamic recovery).

The features of the mechanism of plastic deformation and the properties of individual dislocations cause the formation of dislocation complexes in polycrystalline grains whose properties differ significantly from the properties of individual dislocations. Collective properties manifest themselves at high dislocation densities, when the specific forces of their interaction become greater than the applied stresses and the independent displacements of individual dislocations become impossible [71, 72].

Of particular importance from the point of view of plastic behaviour, deformability and the formation of plastic properties are dislocation ensembles called *flat clusters* and *dislocation walls*. A planar dislocation cluster of edge dislocations is most often the result of the emission of a series of dislocation loops by the Frank–Read source (Fig. 3.9), provided that the leading dislocation stops at some barrier, for example near the grain boundary (Fig. 3.12).

A flat cluster acts as a stress concentrator. The stress at its tip (in the region of the leading dislocation 1) is equal to

$$\tau' = n\tau, \tag{3.15}$$

where τ is the effective (average) stress: n is the number of

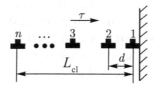

Fig. 3.12. Dislocation distribution in a flat cluster near the barrier.

Fundamentals of the Physics of Strength and Plasticity

dislocations in the cluster. In this connection, this dislocation ensemble can be regarded as a *superdislocation* with the Burgers vector **B** = n**b**. For a certain number of dislocations in the cluster, as follows from (3.15), τ' can reach the value of theoretical strength and the nucleation of a submicrocrack will become energetically favourable, which will lead to relaxation of τ'. The number of possible dislocations on the length L_{cl} for a specific τ is

$$n = \pi L_{cl} \tau k / Gb, \tag{3.16}$$

where $k = 1$ for an edge dislocation and $k = 1-v$ for a screw dislocation, where v is the Poisson coefficient.

The distance between the leading dislocation and the second dislocation is estimated by the equation

$$d \approx \frac{Gb}{\pi k n \tau}. \tag{3.17}$$

Stress relaxation τ' can also occur during the climb of the leading dislocations, which reduces the probability of the submicrocrack nucleation. With increasing temperature, the probability of climb and its speed increase. Consequently, the probability of the emergence of the submicrocrack and, consequently, fracture is reduced. This is one of the reasons for the increase in ductility of metals with increasing temperature.

At high dislocation densities, their total energy is high and is the driving force of their spontaneous climb (the temperature must be high enough) and aligning them in the walls. In the process of climb and aligning into the walls, some of the dislocations annihilate. The dislocation wall consists of disposed edge dislocations located one below the other at equal distances d (Fig. 3.13) [72].

In contrast to the accumulation and chaotic distribution of dislocations of the same density, the dislocation wall has a smaller intrinsic energy, since at a distance from the wall larger than d the components of the stress tensor from its various dislocations cancel each other. The lowering of the energy of the dislocation system also determines the spontaneity (spontaneous flow) of the formation of dislocation walls. The resulting wall from the edge dislocations creates a misorientation in the lattice of the quantity $\varphi = b/d$ (Fig. 3.13) and is the *tilt sub-boundary*. The wall of screw dislocations has the form of a wall from the dislocations of two

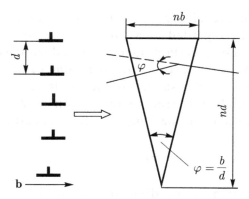

Fig. 3.13. The dislocation wall and its misorientation φ of the crystal.

systems and twists the lattice around an axis perpendicular to the sub-boundary.

The recovery associated with the formation in the polycrystalline grains of a network of sub-boundaries and, accordingly, the subgrains, is called *polygonization* and occurs in the temperature range $0.6 T_m \geq T \geq 0.2 T_m$ (T_m is the melting point). The plastic deformation, carried out at given temperatures and, therefore, under the conditions of dynamic polygonization, is called *warm*. The deformation at $T < 0.2 T_m$ is characterized by intensive hardening and is called *cold*. At $T > 0.6 T_m$ the deformation proceeds under conditions of *dynamic recrystallization* and is called *hot*. Recrystallization is the most efficient recovery process. The mobile boundaries of the recrystallizing grains absorb dislocations (dislocations are expended on the formation of new grain boundaries). *Hot plastic deformation proceeds under the conditions of competition between hardening processes (increasing the density of stationary dislocations with increasing deformation) and recovery (decrease in the density of stationary dislocations*, mainly due to recrystallization). This determines a sufficiently high sensitivity of the flow stress to the strain rate and temperature. At the temperatures of warm and hot deformation, the metals exbihit, except elasticity and plasticity, also ductility – the dependence of σ on $\dot{\varepsilon}$.

If the grain size in metals is of the order of 1.0 μm, then at the hot deformation temperatures there is an interval of strain rates (usually 10^{-3}–10^{-1} s^{-1}), in which the flow becomes predominantly viscous and the *phenomenon of superplastic deformation* is observed – ultrahigh elongations at low stresses [142–145]. In the superplastic flow, the main contribution to deformation is made by grain-boundary slip –

Fundamentals of the Physics of Strength and Plasticity 97

the sliding of the grain groups relative to each other along common boundaries. In this mechanism, an important role is played by diffusion accommodation. Therefore, the deformation is characterized by a high sensitivity of σ to $\dot{\varepsilon}$.

The basic equations of the physical theory of plastic deformation, which relate the macrocharacteristics of deformation (σ, ε and $\dot{\varepsilon}$) to the characteristics of the structure, are obtained as follows. If an edge dislocation passes through the crystal element through the slip plane $ABCD$ along its entire length l to form a step on the left-hand side (Fig. 3.14), then, according to its definition, the plastic deformation of the crystal will be $\varepsilon_1 = b/l$. When n dislocations run, $\varepsilon = nb/l$. Because of the presence of barriers in real crystals, the dislocations travel mean free path $\lambda < l$. Therefore, the deformation of the crystal element will be l/λ times less, that is,

$$\varepsilon = \frac{nb}{l} \cdot \frac{\lambda}{l} = \rho_g b \lambda, \qquad (3.18)$$

where $\rho_g = n/l^2$ is the density of dislocations contributing to deformation.

If the deformation is carried out at a constant speed (ρ_g = const), differentiation of (3.18) with respect to time yields an expression for it, which relates the rate of plastic deformation to the dislocation flow – the Orowan equation:

$$d\varepsilon/dt = \dot{\varepsilon} = \rho_g b v, \qquad (3.19)$$

where v is the average slip velocity of dislocations, which, because of the presence of barriers, is much smaller than the speed of sound c in the metal.

Slip of dislocations with the overcoming of small-scale barriers is a thermally activated process and, by analogy with equation (3.14), for v it is possible to write down

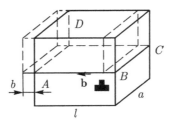

Fig. 3.14. The run of one edge dislocation with the Burgers vector **b** in the elements of the crystal along the slip plane $ABCD$.

$$v = v_0 \exp\left(-\frac{U(\sigma)}{kT}\right). \qquad (3.20)$$

Taking (3.20) into account (3.19) takes the form

$$\dot{\varepsilon} = \rho_g b v_0 \exp\left(-\frac{U(\sigma)}{kT}\right) = \dot{\varepsilon}_0 \exp\left(-\frac{U(\sigma)}{kT}\right), \qquad (3.21)$$

where $\dot{\varepsilon}_0 = \rho_g b v_0$.

An equation of the form (3.21) was first proposed to describe the plastic behaviour of metals at elevated temperatures by Becker [107, 108], by analogy with the Arrhenius equation for the rate of chemical reaction under equilibrium conditions.

If (3.21) is taken as $\dot{\varepsilon} = \dot{\varepsilon}_0 \exp\left[-(U - \tau V)/kT\right]$, then, taking into account $\tau = \sigma/\sqrt{3}$, an expression for the flow stress as a function of temperature and strain rate has the form:

$$\sigma = \sqrt{3}\left[U - kT \ln(\dot{\varepsilon}_0/\dot{\varepsilon})\right]/V. \qquad (3.22)$$

It follows from (3.22) that the flow stress includes the component $\sqrt{3}U/V$ (the thermal component σ) independent of T and $\dot{\varepsilon}$ и and the component $\sqrt{3}kT \ln(\dot{\varepsilon}_0/\dot{\varepsilon})/V$, which depends on T and $\dot{\varepsilon}$ (the thermal component σ).

The equations (3.18), (3.19), (3.21) together with the equations (3.9) and (3.13) form the basis of the physical theory of plastic deformation. Here they play the same role as the defining relations (1.26), (1.28) and (1.29) and (1.35) in the mechanics of plastic deformation.

3.2. Theoretical description of plastic deformation

3.2.1. Multilevel character of plastic deformation

Before proceeding to an exposition of the approach to the theoretical description of plastic deformation in the physics of strength and plasticity, it is necessary to give a brief analysis of the results concerning the concept of the multilevel nature of plastic deformation and fracture of metals that has been intensively developed in the last three decades [35, 71, 148, 149]. Following the authors of the concept, we will briefly consider its essence.

One of its main concepts is the notion of a structural element of deformation, by which is meant a region bounded by a closed surface in a deformable polycrystal, which has a characteristic size of order $\rho^{-1/2}$ and has its own field of long-range stresses.

Real physical objects corresponding to the structural elements of deformation are *cells, fragments, subgrains* and *grains* (crystallites). It is believed that the cause of the formation of these structures in the development of plastic deformation is its heterogeneity, which causes the occurrence of internal stresses in the lattice, including local moments of forces [149]. Hence the following fundamental concepts of the concept are the concepts of the orientational instability of the lattice and rotational (rotational) modes of plastic deformation, whose carriers are disclinations. A partial disclination is a dislocation wall with a length less than the transverse grain size – a discontinuous dislocation boundary (Fig. 3.15) [150]. It leads to misorientation of the lattice by an angle $\varphi = b/d$ (Fig. 3.13).

The dipole of partial disclinations causes a shift of one part of the crystal relative to the other, that is, it is equivalent to a slip band. The nucleation and propagation along the crystal (crystallite) by separation of dislocations along the sign and the attachment to partial walls of the dipole of partial disclinations is a mechanism for polygonization (the appearance of subgrains) and, simultaneously, a mechanism for separating the structural element of deformation and the second deformation level with respect to the atomic–dislocation level. This level has a characteristic size of 0.2–1.0 µm and is called *mesoscopic* [71, 149]. Inside this region, the process of deformation is considered on the basis of classical dislocation concepts, outside it – on the basis of concepts of the rotational modes of plasticity.

The basic assumptions of the concept are reduced to the following [150, 151].

1. Slip of dislocations creates an inhomogeneous deformation in the polycrystalline grains, which is a consequence of the failure of the Polanyi–Taylor scheme (simultaneous operation of five

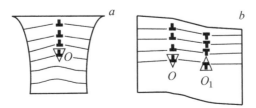

Fig. 3.15. Disclinations: a – partial disclination; b – dipole of partial disclinations.

sliding systems to preserve the continuity of microvolumes). The inhomogeneity of the deformation leads to the appearance of special defects at the boundaries – *butt disclinations* possessing strong internal stress fields. When the last critical values are reached, their relaxation takes place by the collective movement of dislocations, which leads to the rotation of microvolumes as whole. Rotation carriers are *disclination–dislocation complexes*, and the deformation reaches a higher structural level. Rotational modes of deformation come into operation for large (developed) plastic strains, when $\varepsilon > 0.2$.

2. The displacement of the structural elements of deformation as integers, along with dislocation movements in their volumes, contributes to the plastic deformation. The proof is, among other things, superplastic deformation, in which the main contribution to deformation is made by *grain boundary sliding* – gliding of fine (0.3–10 μm) grains relative to each other as whole [152, 153].

3. An adequate description of the large plastic strains of metals within the framework of classical dislocation representations is impossible, it is necessary to take into account the rotational modes of plasticity, that is, the multilevel character of the deformation.

4. New approaches and methods of description are proposed [35, 151].

The content of the physics of developed plastic deformation ($\varepsilon > 0.2$) is elaborated and described in the most detailed manner by V.V. Rybin [71, 150]. He gives a picture of the evolution of the microstructure into the mesostructure, that is, the mechanism of fragmentation – the dispersion of the structure in the process of developed plastic deformation.

In the mechanics of plastic deformation, the subdivision of deformation into *small* ($\varepsilon \leq 0.1$) and *large (developed, finite)* ($\varepsilon > 0.1$) was assumed long ago (see Chapter 1). In mechanics, this subdivision is justified by the following circumstances. In the interval of small strains, the relative strains are equal to the true strains. For small strains, the Cauchy linear tensor of small deformation can be used as their quantitative measure instead of the non-linear Green tensor of finite strains. Small strains with good accuracy can be considered monotonic, loading within small strains is simple. The boundary plasticity problems are solved using the deformation theory of small elastoplastic strains [5].

The modern technology of metal working with pressure tends to use ever higher rates of plastic deformation. In multistage cold volume forging, as already noted, the accumulated intensity of deformation reaches 2–4 or more [33]. In the processes of plastic structure formation – 6–12 and more [34, 94, 154]. Strains of this magnitude in the literature on plastic structure formation have been called intensive, although no definition has been given to this concept, that is, it has not yet received any physical or mechanical justification. In this connection, the concept of a multilevel character of the plastic deformation of metals acquires special significance.

Within the framework of this concept, the gradation of deformation into a small and developed one gets a physical justification. The physics of small deformation is the pattern of evolution of a microstructure determined by microdefects (point, linear, planar). The physics of developed plastic deformation is the pattern of the evolution of the mesostructure determined by mesodefects (partial disclinations, disclinational dipoles, butt disclinations, fragments) [150].

In the above-described concept of the physics of developed plastic deformation, the fragmentation mechanism, that is, the evolution of the initial microstructure into a microcrystalline one, in which the boundaries between the fragments are large-angle, and the grain fragments have a mesoscopic size $d = 0.2–1.0$ μm, has a theoretical and, most importantly, experimental substantiation [150]. However, in this picture, the argument leading to the possibility of the existence of rotational modes of plastic deformation is, in our opinion, debatable. As an alternative, not claiming proven truth, we give the following argument.

The author of [150], concisely expounding the essence of the developed physics of developed plastic deformation, in establishing the rotational modes and, correspondingly, the multilevel nature of the deformation, repels the notion of the *Taylor rotation of the grain* (crystal). If an unoccupied crystal *ABCD* under stress σ is deformed by ε by a laminar flow of dislocations moving along parallel sliding systems, then the crystal, being unrestricted, rotates relative to the initial position by an angle θ (Fig. 3.16 *a*). This Taylor rotation is not accompanied by a reorientation of the crystal lattice.

In a real polycrystal, the grains are constrained by neighbours. Therefore, they can be plastically deformed and rotated only isomorphically with respect to macroscopic shaping and macroscopic rotation of the body. The realization of the second condition is attributed by the author [150] to the necessity of introducing, in addition to the Taylor rotation (Fig. 3.16 *a*), reactive rotation of the

102 *Physico-Mathematical Theory of High Irreversible Strains*

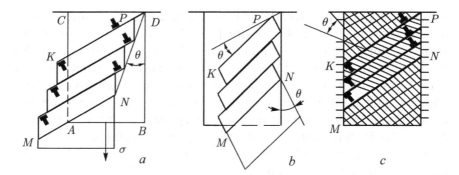

Fig. 3.16. Schemes of deformation of the crystal (grain) by sliding dislocations: *a* – Taylor rotation of a part of the crystal *MNKP*; *b* – reactive rotation – rotational mode of plasticity; *c* – mechanism of elastic reorientation of the lattice as a result of the incomplete run of an ensemble of edge dislocations along parallel slip planes.

grain volume *MNKP* by an angle θ, as shown in Fig. 3.16 *b*. The reactive rotation is associated with the reorientation of the lattice in the *MNKP* volume, i.e., it changes the metric of a given volume – it plasticizes the volume by a shear by an amount $\varepsilon = \tan \theta$. And this is considered as a deformation, additional to the one that the volume received due to the run of dislocations.

The author of [150] states: "Rotational modes of plasticity convert material plastic rotations with an invariant lattice into a reorientation of the crystal lattice of the grain."

However, the Taylor rotation in an invariant lattice under the deformation of a polycrystal is never realized, since the grains in it are constrained by the neighbours. Therefore, for example, with the expansion of dislocation loops in parallel planes of the grain volume *MNKP* (Fig. 3.16 *c*), the dislocations stop either at barriers, for example near grain boundaries, or from the action of internal stresses with the formation of a dipole of partial disclinations. Dislocations have travelled a certain path λ along the crystal, which means that its plastic deformation has taken place with the value $\varepsilon = \rho b \lambda$. Two dislocation walls, *MK* and *NP*, rotated (shifted) the *MNKP* volume relative to the environment by an angle $\theta = b/d$ (see Fig. 3.13). In this case, the shear strain $\varepsilon = \tan \theta = \rho b \lambda$. It turns out that the deformation of the constrained volume *MNKP* can be described as *slip of dislocations and as a rotation of the volume as a whole (rotation)*. Apparently, this is the fundamental property of plastic deformation of crystals, due to the fundamental property of its carriers – the *translational–rotational dualism of dislocations*.

One edge dislocation in the volume of the crystal already causes a perturbation of the crystal lattice of the rotational type.

The stresses induced by the elastic rotation of the *MNKP* volume relative to the environment caused by the slip of dislocations in parallel planes (Fig. 3.16, *c*) relax as the deformation develops, and the *MNKP* volume evolves to the fragment, as described in [150].

Apparently, the rotation of structural deformation elements at the stage of developed plastic deformation is an accommodative process. Otherwise, an adequate description of the $\sigma(\varepsilon)$ dependences of metals to large values of ε, based only on the inclusion of slip dislocations (without involving rotational modes), for example, equations of the type $\sigma = \alpha mb\sqrt{\rho}$, $\rho = \varepsilon/b\lambda$ and $\sigma = \alpha mDb\sqrt{\varepsilon/b\lambda}$, would be impossible, which is not true [155].

The application by the authors of the concept of the mechanism of structural superplasticity – the sliding of small grains relative to each other as whole along a common boundary – is, in our view, incorrect when trying to prove the independent contribution of rotations of structural elements as whole to plastic deformation. Superplasticity is the state of a microcrystalline metal, i.e., a metal in which the mesostructural level is preset initially. In terms of the physics of developed plastic deformation, this is a fragmented structure. Consequently, superplastic deformation is the *developed deformation*.

Like the deformation of a fragmented structure under superplastic deformation, the Polanyi–Taylor scheme is not satisfied. Grains with a linear size of 0.2–1.0 μm remain equiaxed to very large strains. We note here that the stage of developed plastic deformation is characterized by the process of fragmentation (evolution of the mesostructure). Strains that do not fulfill the Polanyi–Taylor principle (the process of fragmentation has come to an end) [150], can be called intensive, although their mechanism is, in general, still unclear.

It has been experimentally proved that the grain boundary gliding during superplasticity introduces a major (up to 80%) contribution to plastic deformation. But gliding is a mechanism of sliding when the groups of grains (from a few tens to hundreds) slide as a whole relative to each other along common boundaries lying approximately in one plane. This is an experimentally proven fact [156], which was theoretically predicted in 1985 [157].

The question of the mechanism of this cooperated grain-boundary slip of the grain group as a whole structural element remains a discussion. Either it is slip along liquid-like boundaries that have migrated to a given state as a result of interaction with the flow of lattice dislocations [158, 159] or it is grain-bound slip caused by the sliding of grain boundary dislocations stimulated by the interaction of the grain boundaries with the flow of lattice dislocations [160] or sliding, due to the slip of super-grain boundary dislocations with an average Burgers vector whose modulus is equal to the linear grain size and which were called *crystallites* [143–145].

The fact that the superplastic flow can be described by equations analogous to the dislocation plasticity equations [161] once again indicates that grain boundary slip is gliding and not rotation. Consequently, in the case of superplastic deformation, which is observed in materials with a specially prepared structure ($d = 0.2$–1.0 µm), the second level of deformation is related to the *cooperated gliding of the grain groups*. The distance between parallel shear planes is $(10$–$100)\, d$. If we take $d \approx 0.2$ µm, then the linear dimension of the deformation structural element will be 2.0–20.0 µm.

Cooperated grain-boundary slip must have its own elementary carrier. In this role, grain boundary dislocations or super-grain boundary dislocations, crystallite, can act. The slip of a crystallite is equivalent to slipping of the grain along a liquid-like boundary. Since the grain boundary and, especially, the glide plane passing through the set of boundaries are not crystallographically smooth, the slip of the crystallite should be accompanied by powerful diffusion flows that are an integral part of the gliding mechanism.

It is not difficult to imagine that in the process of multiple cooperated grain-boundary slip rotations occur both in whole groups of grains and in individual grains. These rotations are observed during superplastic deformation. They, as in the case of deformation of coarse-grained metals at the stage of developed deformation, are, most likely, accommodative in nature. It has been experimentally proved that the total deformation at superplasticity is $\varepsilon' = \varepsilon_g + \varepsilon_{dc} + \varepsilon_{gbs}$, where ε_g is the contribution to the deformation of dislocation gliding in grains; ε_{dc} is the strain deformation of diffusion creep and ε_{gbs} is the deformation by the mechanism of grain-boundary slip. There is no contribution from the rotations.

Based on the foregoing, we can conclude that the second level of deformation (after microstructural) comes into operation after the fragmentation process is completed at the stage of deformed deformation. After this stage, the deformation is effected by a mechanism that practically does not lead to a change in the structure, that is, without the Polanyi–Taylor principle predicting an isomorphous change in the sizes and shapes of all internal volumes of a uniformly deformed macroscopic body [150]. The sphere selected in the sample under monotonic deformation is transformed into an ellipsoid.

Such deformation should proceed at a constant stress. In the case of superplastic deformation, the mechanism that does not change the microstructure over a significant strain is the above-described cooperated grain-boundary slip. Therefore, superplastic deformation is an example of the two-level nature of plastic deformation. The main contribution to deformation is made by the mechanism of the second level – cooperated grain boundary slip. The linear dimension of the deformation structural element is of the order of 2–20 μm. The mechanism of the second level is provided by the mechanisms of the first (microstructural) level – intragrain dislocation gliding and diffusion creep.

Thus, with superplasticity, a special deformation mechanism works and deformation occurs at a constant stress, if T and $\dot{\varepsilon} = \text{const}$. This indicates that the material does not accumulate elastic energy. Consequently, all work of deformation is dissipated into thermal energy. This is possible only if the deformation mechanism is controlled by diffusion.

From the point of view of synergetics [162], the superplastic state of the fine-crystalline material can be explained as follows. Lattice dislocations, leaving the boundaries of grains, bring additional atoms and elastic energy, that is, the boundaries, as a system, exchange with matter and energy with the environment, thus being an open system. At a certain intensity of this process (the critical strain rate), the thin grain boundary layer of the material passes into a special excited state (undergoes a kinetic phase transition) [163–168], in which the grain-boundary diffusion coefficient increases sharply, and the boundary system passes to a new stable state, leading to the appearance of a stationary process of cooperated grain-boundary slip. In the deformed finely crystalline material, a coherent interaction of parts of the system occurs, i.e., a dissipative structure is formed [162]. Only such an explanation of superplasticity allows us to speak of it as a special excited state of a fine-crystalline metal.

With this approach, the superplasticity curve lg σ(lg $\dot{\varepsilon}$) can be interpreted as follows (Fig. 3.17). At an optimum temperature (the rate sensitivity coefficient has the maximum value $m_{\max}$ = $\partial \lg \sigma / \partial \ln \dot{\varepsilon}$), the deformation of the microcrystalline metal for very small $\dot{\varepsilon} < 10^{-5} - 10^{-6} \text{s}^{-1}$ is realized by the Coble diffusion creep (DC) mechanism [143]. Diffusion mass transfer proceeds along grain boundaries, and σ is directly proportional to $\dot{\varepsilon}$:

$$\sigma = \frac{kTd^3}{B\Omega w D_{gb}} \cdot \dot{\varepsilon}, \qquad (3.23)$$

where B is a numerical constant, depending on the shape of the grains; d is the linear grain size; Ω is the atomic volume; w is the width of grain boundaries; D_{gb} is the coefficient of grain-boundary diffusion.

From the energy point of view, the diffusion creep is the most 'low-lying' deformation mechanism. Therefore, by analogy with the basic state of the physical system, deformation by the diffusion creep mechanism was called the *basic deformation state*. In contrast to the ground state of the system, which is equilibrium, the basic deformation state is a non-equilibrium stationary process.

With increasing $\dot{\varepsilon}$, according to (3.23), σ increases, and at some value the tangential stresses in grains become equal to the critical value τ_{cr}. Then, according to the Schmid law (3.6), the mechanism of intragrain dislocation sliding (IDG) comes into operation. This mechanism, in comparison with DC, is a more effective mechanism of mass transfer. Therefore, the dependence $\sigma(\dot{\varepsilon})$ becomes non-linear, and for a certain intensity of the lattice dislocation flow falling on the boundary, corresponding to the critical value of the kinetic parameter $\dot{\varepsilon}_*$, as noted above, the grain boundary system becomes a special excited state, which causes the co-operative grain-boundary slip of the grain groups as a whole. In terms of the quasiparticle method, this excited state, by analogy with the quasiparticle excitations of the ground state, was interpreted as the excitation of the basic deformation state (laminar diffusion flow) and is called the crystallite. Then, the crystallite should be regarded as the excitation of the basic deformation, that is, microstructural level of deformation, and the entry into operation of a mechanism of a higher structural level, which is an unfinished cooperated shift of the grain groups as integers relative to each other along common boundaries lying approximately in one plane, on background of powerful diffusion flows along these boundaries – diffusion creep. In terms of the theory of dissipative structures, the value of $\dot{\varepsilon}_*$ in Fig. 3.17 is the bifurcation point. Curve *1* (dashed curve) is an unstable thermodynamic branch. Curve *2* is a new stable state (dissipative structure) – superplasticity.

At the bifurcation point $\dot{\varepsilon}_*$ the main deformation state of the material is excited, therefore, with further increase of the kinetic parameter $\dot{\varepsilon}$, its energy must remain constant, since its increase with increasing $\dot{\varepsilon}$ due to the increase of σ in (3.23) should be spent on the formation of excitations, i.e., the stress σ_D for $\dot{\varepsilon} > \dot{\varepsilon}_*$ must remain constant (line 3 in Fig. 3.17). This is possible if, for $\dot{\varepsilon} > \dot{\varepsilon}_*$ in (3.23), the grain-boundary diffusion coefficient is proportional to the plastic strain rate $D_{gb} = \chi\dot{\varepsilon}$. An experimental verification of this prediction of the theory will be one of the proofs of the above approach to explaining the phenomenon of superplasticity of metals.

Equation (3.23) takes the form

$$\sigma_b = \frac{kTd^3}{B\Omega w \chi}. \tag{3.24}$$

For a particular material at constant temperature in (3.24), σ_b = const in a wide velocity range of superplasticity and is the stress of the basic deformation state – the first structural level of deformation (background stress). The total stress will be equal to

$$\sigma = \sigma_b + \sigma_{cr}(\dot{\varepsilon}), \tag{3.25}$$

where σ_{cr} is the stress due to excitations (gliding and interaction of crystallites) – the stress of the second structural level of deformation.

Assuming that the scalar density of crystallites is proportional to the rate of superplastic deformation $\rho_{cr} = C\dot{\varepsilon}$ and the rates at which they arise, being classical quasiparticles, obey the normal distribution law with the average (optimal, when $m = d \lg/d\sigma g\dot{\varepsilon} = m_{max}$) strain rate $\dot{\varepsilon}_0$ and the variance S_0, we obtain

$$C(\dot{\varepsilon}) = \frac{\rho_{cr}^{max}}{\sqrt{2\pi}} \exp\left[-\frac{1}{2}\left(\frac{\dot{\varepsilon} - \dot{\varepsilon}_0}{S_0}\right)^2\right]. \tag{3.26}$$

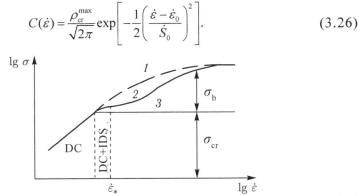

Fig. 3.17. The representation of superplasticity as a special non-equilibrium excited state – dissipative structure.

Since the crystallites differ from the dislocations by the Burgers vector, by analogy with (3.9) we obtain

$$\sigma_{cr}(\dot{\varepsilon}) = \alpha' b' G' \sqrt{\rho_{cr}}. \qquad (3.27)$$

Thus, superplastic deformation is a classic example of the two-level nature of plastic deformation of metals. In this case, the deformation mechanism of the second structural level has a shear character.

3.2.2. Structure and properties of metals with developed and intense plastic strains

It has been experimentally established that in the case of deformation of a metal with an initial coarse-grained microstructure, the fragmentation process and, consequently, the stage of developed deformation terminate when a fragmented structure with an average linear size of 0.2 µm fragments is formed in the entire volume of the deformed metal [150]. This occurs when the intensity of the accumulated plastic deformation is about 3–4 (Fig. 3.18) [150].

Figure 3.19 shows the flow stresses of three alloys as a function of deformation when the latter is varied to six units. They were constructed using the following procedure.

Cylindrical billets with a diameter of 15 mm and a length of 80 mm were deformed with reproduction of the original shape and dimensions according to the 'soft' pressing scheme, as shown in Fig. 3.20 [169, 199].

In one processing cycle, in which the workpiece was deformed by extrusion with simultaneous upsetting (Fig. 3.20 a, b), the average strain intensity was $\varepsilon = 4 \ln (D/d) = 0.9$. The strain accumulated in n processing cycles is $n\varepsilon$. The billets were deformed

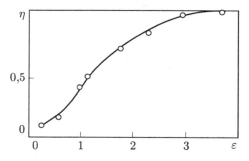

Fig. 3.18. Dependence of the fraction η of the volume affected by fragmentation on the true (logarithmic) strain for Armco iron, deformed by drawing at $T = 300$ K.

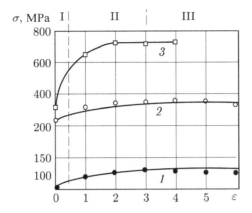

Fig. 3.19. Diagrams of deformation of aluminum AD1 (*1*), copper M1 (*2*) and steel 08kp (*3*) at a temperature of 20°C (293 K) over a wide range of strains: I, II and III – the stage of small ($\varepsilon \leq 0{,}2$), developed ($0.2 \leq \varepsilon \leq 2.0$–$4{,}0$) and intensive ($\varepsilon > 2.0$–$4.0$) strains respectively.

at room temperature with different strain rates without disrupting the continuity due to the 'soft' scheme of the stressed state (comprehensive uneven compression).

From the middle of the workpieces, where the deformation proceeded uniformly (Fig. 3.21), standard cylindrical five-fold specimens were prepared for tensile loading and with a diameter of 6 and a height of 8 mm for upsetting, using a standard procedure to determine the yield stress σ_{02}. When constructing the deformation diagrams, the value of σ_{02} was put in correspondence with the strain accumulated by the workpiece during pressing.

It can be seen (Fig. 3.19) that the hardening of three different metals occurs approximately in the strain range from $\varepsilon = 0$ to $\varepsilon = 2.0$–4.0. Further, the $\sigma(\varepsilon)$ dependences reach the plateau and the flow stress remains constant up to very large strains. Even an insignificant decrease in σ, as is usually the case with hot deformation (Fig. 1.6), ca be observed. In this case, the difference in the values of σ_{02}, determined by stretching and upsetting, was within the range of the scatter of the σ_{02} values from sample to sample, that is, the anisotropy of the properties does not arise under intense alternating strains.

Note that the described character of the $\sigma(\varepsilon)$ dependences is not new. Analogous dependences were obtained earlier on other alloys [57]. At present, on the basis of the multilevel nature of plastic deformation accumulated during the investigation, it can be given a physical interpretation.

110 *Physico-Mathematical Theory of High Irreversible Strains*

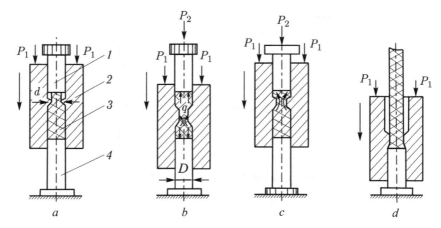

Fig. 3.20. Scheme of the process of 'hourglass' metal pressing: *1, 4* – punch; *2* – matrix; *3* – workpiece blank (sample)

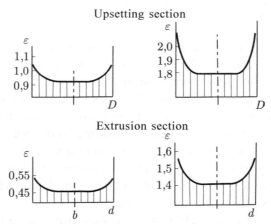

Fig. 3.21. Distribution of true strain along the workpiece cross-section: *a* – one processing cycle; *b* – two processing cycles.

Investigation of the microstructure of the blanks showed that at a strain value of 2–4, the size of the grains decreases from 1.5–2.0 to 0.7–0.8 µm. With a further increase in strain, the grain size does not change and the grains remain equiaxial [169, 170]. The lack of accumulation of elastic energy in metals under deformation at $\varepsilon > 2$–4 (σ = const) and the stable state of the structure indicate a change in the deformation mechanism. At the strain of $\varepsilon = 2$–4, the process of forming grains of microcrystalline size (the process of fragmentation) ends and, it can be assumed, a new deformation mechanism enters into operation, associated with its emergence to a new structural level. To date, only one mechanism of this level is known, which operates in the presence of a microcrystalline structure, contributes independently to deformation and does not lead to hardening – it is a cooperated grain-boundary slip in superplasticity. The possibility of its operation under cold deformation has not been experimentally proved, although it is known that in microcrystalline materials, even under static conditions, the grain-boundary diffusion coefficient is 5–6 orders of magnitude higher than in the corresponding coarse-grained materials [171]. In the presence of shear stresses applied to the boundaries (under deformation conditions), the diffusion coefficients may increase by several orders of magnitude due to the appearance of strongly excited atom–vacancy states [166, 167]; to the experimental proof of the existence of this is provided in [168].

The above results allow us to refine the boundaries of the intervals of small, developed and intensive strains and to specify the formulation of the problem of their theoretical description. From the point of view of physics and mechanics, the division of plastic deformation into small ($\varepsilon \leq 0.2$), developed (large, finite) ($0.2 < \varepsilon \leq (2.0–4.0)$) and intensive ($\varepsilon > (2.0–4.0)$) is substantiated. In the range of developed deformation there are fairly fundamental changes in the structure of the metal due to the process of fragmentation – the formation and evolution of the mesostructure. The main content of this process is the dispersion of the structure, the leading mechanism is the motion of partial disclinations, leading to rotation of submicroscopic volumes in grains [150].

The stage of *intense deformation* is observed at intensities of the accumulated deformation of $\varepsilon > (2.0–4.0)$. The grounds for distinguishing this stage are: from the point of view of the mechanics of plastic deformation – the state of ideal plasticity, for which σ does not depend on ε (Fig. 3.19); from the point of view of the physics of plasticity – the work of a specific deformation mechanism, in which the mesostructure does not change. In light of the justified classification of strain by size, the process of fragmentation (grain dispersing), which is the basis of the technology of plastic structure formation [154], proceeds and ends mainly at the stage of developed deformation of $0.2 \leq \varepsilon \leq (2.0–4.0)$. At the stage of intense strain $\varepsilon > (2.0–4.0)$ under traditional deformation schemes, the microstructure is stable up to deformation of 6–12 and more.

The postulate of ideal plasticity and the strengthened postulate of ideal plasticity are formulated in [57] and within the framework of mechanics. The first postulate states that, starting from a certain moment of deformation, the materials become perfectly plastic, i.e., when $\varepsilon \geq \varepsilon^*$ σ = const. In this case, ε^* and σ depend on the deformation history for $\varepsilon < \varepsilon^*$.

The second postulate suggests that in addition to the loading surface, for each material in the stress space there is a fixed (independent of the loading history) limiting surface of ideal plasticity, corresponding to a homogeneous and isotropic material. Starting from a certain moment of deformation (at $\varepsilon = \varepsilon^*$), the end of the stress vector and, consequently, the loading surface touch this surface, and for $\varepsilon \geq \varepsilon^*$ the vector σ_{ij} moves along it, that is, the material behaves as perfectly plastic.

The above results can be considered a physical justification for these postulates.

However, it should be noted that there are experimental data that indicate a different (than in Fig. 3.19) dependence of the characteristics of the mechanical properties on the degree of deformation in the interval of intense deformation. The authors of [172] investigated the effect of the degree of cold deformation under torsion in the conditions of quasi-hydrostatic pressure (torsion on Bridgman anvils) and rolling on the structure and microhardness of single crystals of copper, nickel, and KhN77TYuR alloy. Torsion on the Bridgman anvils was applied to deform flat samples with an initial thickness of 2.0 mm to a final thickness of 0.05 mm. The obtained microhardness dependences on the true deformation are shown in Fig. 3.22.

Up to the strain $\varepsilon \approx 4.0$, the character of the $H_\mu(\varepsilon)$ dependences is completely identical to the $\sigma(\varepsilon)$ dependences in Fig. 3.19. Beginning at $\varepsilon \approx 4.0$, the materials are again strengthened with access to a new plateau at $\varepsilon \approx 5.0$. Secondary hardening at $\varepsilon > 4.0$ was also observed during cold deformation by rolling and drawing of pure polycrystalline Pt, Ni, $Pt_{90}Ir_{10}$, $Pt_{93}Rh_3Rd_4$ (obtained by vacuum melting) [134].

Studies of the structure of materials have shown that at $\varepsilon > 4.0$ the structure is further dispersed from 0.5 to 0.1–0.2 μm and an increase in the misorientation between the crystallites takes place. In this case, the microcrystallites remained equiaxial and there was practically no crystallographic texture. This indicates the work of the deformation mechanism of the second level.

In both studies, deformation was applied to very thin samples. Therefore, the stress state in the centre of plastic deformation was characterized by a high value of the positive hydrostatic pressure. The pressure of 6.0 GPa can to some extent suppress the diffusion process

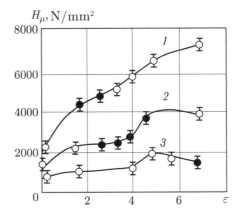

Fig. 3.22. The microhardness of the KhN77YuR alloy (*1*), nickel (*2*) and copper (*3*), depending on the degree of plastic deformation in torsion (○) and rolling (•).

of relaxation of moment stresses, competing with fragmentation, and the fragmentation will continue.

In the industry the pressure treatment of metals is carried out by deformation procedures and workpieces in which it is impossible to achieve hydrostatic pressures of this magnitude. Nevertheless, a natural and important question arises about the limit of dispersion of the structure at the stages of developed and intensive deformation.

Recently, studies have been carried out of the structure and properties of microcrystalline (ultrafine-grained) metals, obtained, as their authors believe, by severe plastic deformation [154, 172–175]. The problem of obtaining nanostructured (grain size $d \leq 100$ nm) bulk semifinished products by the method of severe plastic deformation is discussed [176].

Is it possible using a process of plastic structure formation to obtain a nanostructured state that provides unique properties of structural materials? In [177], experimental data on the minimum grain size d^* obtained in the deformation of metals by the method of equal-channel angular pressing (ECAP) and the torsion method under quasi-hydrostatic pressure conditions, which are given in Table 3.1, are collected. It can be seen that the nanostructured state ($d^* = 60$ nm) was obtained once during the torsion of molybdenum in the Bridgman anvils. This method produces samples with a diameter of 5–10 mm and a thickness of 0.05–0.5 mm [173]. When torsion machining a metal flake with a thickness of the order of 0.1 mm, in addition to the effect of pressure on diffusion relaxation noted above there may also occur mechanical dispersion (grinding) followed by the consolidation (welding) of highly activated particles under high pressure. Therefore, this result is not indicative. In addition, a sample of this size is of no interest to industry.

From Table 3.1 it follows that the minimum grain size $d^* = 0.25$–0.30 μm obtained at room temperature is the most stable. With a slight increase in the treatment temperature, d^* increases quite sharply.

The authors of [177] obtained a theoretical expression for d^* in the form

$$d^* = \left(\frac{\omega^* \delta^* D^*}{A_1 \xi \dot\varepsilon} \frac{G\Omega}{kT} \right)^{1/3}, \qquad (3.28)$$

where ω^* is the critical power of the butt disclinations, at which emission starts from the junctions of broken dislocation walls (the mechanism of fragmentation — grinding of the grains starts); δ^* is the width of the boundary; D^* is the grain-boundary diffusion coefficient; G is the shear modulus; Ω is the atomic volume; A_1 is the coefficient; ξ is the coefficient of homogeneity of plastic deformation; $\dot\varepsilon$ is the plastic strain rate; k is the Boltzmann constant; T is the temperature.

Information on d^*, collected from various sources in [177], will be supplemented with new results from the recent papers [178–180], given in Table 3.2[1].

It follows from the table that only submicrocrystalline non-ferrous metals and their alloys with a grain size of 0.2 μm or more are obtained by different deformation methods.

The authors of Ref. [177] explain the existence of the limit d^* in the framework of the physics of developed deformation on reaching d^* and the corresponding value D^* of the new mechanism of accommodation of butt disclinations — diffusion accommodation, which is an alternative to the mechanism of their disrupted dislocation wall emission — the mechanism of fragmentation (refining) of the structure. For small fragment sizes, the rate of grain-boundary diffusion, determined in (3.25) by the value of D^*, becomes very high and the fragmentation mechanism ceases to work.

To decrease D^* in the process of deformation and, therefore, in accordance with (3.29), we can reduce d^* only by very high compressive stresses of the spherical tensor in the stress state scheme, which reduce the rate of diffusion. Apparently, this explains

---

[1] In [177–180] the values of d^* are given as those obtained by the authors themselves, as well as those borrowed from other sources.

Fundamentals of the Physics of Strength and Plasticity

Table 3.1. The minimum grain size d^* obtained by processing various metals and alloys by ECAP and torsion on Bridgman anvils [177]

Metal (alloy)	Deformation temperature, K	d^*, μm ECAP	d^*, μm Torsion in Bridgman anvils
Al	293	1.30	
Cu	293	0.30	
Ni	293	0.10	
Fe	293	0.10	0.1; 0.16; 0.4; 0.15
Cr	293		0.1
Mo	308		0.06
Al–0,22Sc–0,15Zr	373	1.20	
Al–0.22Sc–0.15Zr	433	1.80	
Al–1.5Mg–0.22Sc–0.15Zr	373	0.30	
Al–1.5Mg–0.22Sc–0.15Zr	473	0.70	
Al–3Mg–0.22Sc–0.15Zr	373	0.35	
Al–3Mg–0.22Sc–0.15Zr	473	0.80	
5052 (Al–Mg system)	323	0.25	
5052 (Al–Mg system)	373	0.30	
MA2–1	473	1.70	
AZ91 (Mg–Al system)	423	1.30	
AZ91 (Mg–Al system)	293	1.0	
Copper of technical purity M06	293	0.25	
Copper of technical purity M3	293	0.20	
Cu–0.3%Cr	293	0.30	
Cu–0.8%Cr–0.1Zr	293	0.30	
Al–1.5%Mg–0.22%Sc–0.15%Zr	393	0.30	
Al–3%Mg–0.2Sc	293	0.20	
Al–3.0Mg	293	0.27	

the obtaining on the Bridgman anvils material with a torsion with d^* equal to 0.1 μm, when the material being processed undergoes shear deformation under conditions of a very high positive hydrostatic pressure $p = -\sigma_0$, where σ_0 is the stress of the spherical tensor. Such pressures are unattainable under other deformation schemes. Therefore, the formulation of the problem of obtaining nanostructured

Table 3.2. The values of the limiting grain size d^* for different metals and deformation methods

Material	Deformation method	d^*, μm	Source
Al–Mg–Si	Torsion	0.24	[178]
Al99.99%	ECAP	1.0	[179]
Al–1050	ECAP	0.6	[179]
Al–Zr–Fe–Si	ECAP	1.0	[179]
Al96%	ECAP	0.4	[179]
Al–3Mg	ECAP	0.15	[179]
Cu99.953	ECAP	0.2	[180]
Cu99.953	ECAP + rolling	0.15	[180]

bulk semifinished products of large sizes by the method of plastic structure formation has no theoretical or experimental justification at present.

3.2.3. Methods of theoretical description of plastic deformation

In the light of the above classification of plastic deformation in magnitude, which has mechanical and physical basis, it can be assumed that a quantitative theoretical description of small and developed deformation is possible within the framework of dislocation concepts. A fundamentally different mechanism of the mesoscopic level operates in the interval of intensive deformations. However, there are papers devoted to the description of structural superplasticity and deformation mechanisms at the mesolevel, showing that the deformation determined by co-operative grain-boundary slip is accomplished by sliding certain grain-boundary defects and can be described on the basis of equations analogous to the dislocation theory of plasticity equations (3.9), (3.18), (3.19), (3.27) [181, 161].

When processing metal under pressure, plastic shaping is carried out in the intervals of small and developed strains. Therefore, let us consider the methods of the theoretical description of plastic deformation on the basis of dislocation representations.

In the physics of strength and plasticity, deformation models are constructed for a uniaxial stress state. On the basis of analysis and generalization of a wide range of experimentally established regularities of the plastic flow of metallic materials of various classes, it is assumed that in the general case the process of plastic deformation can be described by a system of kinetic differential equations [69]:

$$\dot{\varepsilon} = f_1(\rho, c_k, \sigma); \quad \dot{\rho} = f_2(\dot{\varepsilon}, \rho, c_k, \sigma); \quad \dot{c}_k = f_k(\dot{\varepsilon}, \rho, c_k, \sigma), \quad (3.29)$$

where c_k is the concentration of deformation point defects of the k-th type; the dot above the symbol, as usual, means differentiation with respect to time.

Taking into account that the plastic deformation proceeds under the conditions of competition between the hardening and dynamic recovery processes, for the increments of the deformation defect density with an increase in the degree of deformation by a value of $d\varepsilon$, we can write the following relations [69]

$$d\rho = \frac{\partial \rho}{\partial \varepsilon} d\varepsilon + \frac{\partial \rho}{\partial t} dt; \quad dc_k = \frac{\partial c_k}{\partial \varepsilon} d\varepsilon + \frac{\partial c_k}{\partial t} dt, \quad (3.30)$$

where $\partial\rho/\partial\varepsilon$, $\partial c_k/\partial\varepsilon$ are the intensities of multiplication of dislocations and point defects; $\partial\rho/\partial t$, $\partial c_k/\partial t$ are the rate of annihilation of dislocations and point defects.

The system of equations (3.29) taking into account (3.30) takes the form

$$\dot{\varepsilon} = f_1(\rho, c_k, \sigma), \quad \dot{\rho} = \dot{\varepsilon}\frac{\partial \rho}{\partial \varepsilon} + \frac{\partial \rho}{\partial t}. \quad (3.31)$$

It follows from (3.31) that the problem of the mathematical description of plastic deformation can be reduced to finding and solving the kinetic equation for the dislocation density. The simplest kinetic equation for the case of low dislocation density can be taken in the form [69]

$$\frac{d\rho}{dt} = A_1 \rho, \quad (3.32)$$

where $A_1 = A_1(\sigma)$ is the multiplication factor of dislocations.

During plastic deformation, the density of dislocations increases rather rapidly. Some of them stop at the barriers and become immovable. Fixed dislocations are new barriers. Therefore, beginning with a certain degree of deformation, the rate of increase in the dislocation density slows down, which is neglected by Eq. (3.33). In connection with this circumstance, the concept of the density of dislocations moving at a given time ρ_g was introduced and a system of equations was proposed for the description of creep at σ = const [69]:

$$\frac{d\rho_g}{dt} = A_2\rho_g - A_3\rho_g^2, \quad \dot{\varepsilon} = b\rho_g \upsilon(\sigma), \tag{3.33}$$

where the second equation is the Orowan equation (3.19). It relates the macroscopic characteristic of plastic deformation $\dot{\varepsilon}$ to the characteristics of the structure b, ρ_g, v and is therefore one of the basic equations of the dislocation theory of plasticity; A_2 and A_3 are coefficients depending on σ.

The equations (3.33) already sufficiently clearly demonstrate the tendency of the development of the method of theoretical description of plastic deformation in the direction of the kinetic approach.

Further developing the kinetic approach, R.I. Nigmatulin and N.N. Kholin proposed a kinetic equation for the rate of multiplication of dislocations [182]

$$\frac{d\rho}{dt} = mb\rho v_s(\tau)\psi_s(\tau,\rho) + \Gamma(\tau,\rho) - a_\alpha b\rho v_\alpha(\tau)\psi_\alpha, \tag{3.34}$$

where the first term on the right-hand side describes the change in the density of dislocations as a result of multiple cross slip, the second term is the production of dislocations by Frank–Read sources, and the third is the annihilation of dislocations upon creep.

Using the equations (3.34), these authors proposed theoretical models for describing the processes of superplastic deformation and creep of metals. We note that the authors' approach [182] is not purely kinetic, since the parameters of equation (3.34) are chosen from the condition of a satisfactory description of the experimental deformation curves.

An important achievement in the development of the method of theoretical description of plastic deformation, in our opinion, is the approach proposed and developed by B.A. Grinberg and M.A. Ivanov and co-authors [183–185]. The approach is based on one of the methods of kinetics – kinetic equations of detailed balance [54], which are used to describe the processes of transformation of dislocations of different types. The diagram of dislocation transformations is used for clarity. As the simplest diagrams used to describe the kinetics of the mutual transformation during the plastic deformation of mobile g and stationary s dislocations, the authors proposed the 'petal' and 'ray' diagrams (Fig. 3.23). Points on the diagrams denote dislocations of different types, for example, o – mobile dislocations, • – stable. The arrows on the lines indicate the

Fundamentals of the Physics of Strength and Plasticity

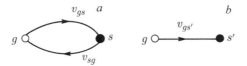

Fig. 3.23. The simplest diagrams of the mutual transformation of mobile and stationary dislocations: *a* – 'petal'; *b* – 'beam'.

directions of dislocation transformations. The diagrams reflect certain physical ideas about micromechanisms of dislocation plasticity. The «petal» diagram can be interpreted as follows.

When the material is loaded by external forces and the yield stress is reached, the dislocation sources generate mobile dislocations *g*. The latter, having traveled a distance equal to the mean free path, stop at obstacles (barriers) of various nature (stationary dislocations of the forest, dispersed particles of the second phases, subgrain boundaries and grains, etc.) and become stationary dislocations *s*. This transformation occurs at a frequency of v_{gs}. The indices with the symbol v show the direction of the transformation $g \to s$.

The stationary dislocations *s*, depending on the nature of the barrier, can bypass it, for example, edge dislocations by the creep mechanism under the action of applied stress and thermal activation, and again become mobile type *g*. This transformation occurs with a frequency v_{sg}. Barriers can be insurmountable for dislocations during the characteristic observation time. Such, for example, are the large-angle grain boundaries at cold deformation. In this case, the process is described by a 'ray' type diagram (Fig. 3.23 *b*), where *s'* are insuperable (impermeable barriers).

Complex diagrams represent different combinations of two protozoa. As an example, Fig. 3.24 shows a diagram which is a series connection of the 'petal' and 'ray' diagrams.

On the basis of the diagrams, the kinetic equations of the detailed balance of dislocation densities are recorded. For the 'petal' diagram (Fig. 3.23 *a*), the equations will have the form

$$\begin{cases} \dfrac{d\rho_g}{dt} = -\rho_g v_{gs} + \rho_s v_{sg} + M, \\ \dfrac{d\rho_s}{dt} = \rho_g v_{gs} - \rho_s v_{sg}, \end{cases} \quad (3.35)$$

where *M* is the power of dislocation sources.

120 *Physico-Mathematical Theory of High Irreversible Strains*

Fig. 3.24. A complex diagram of dislocation transformations.

In the method, the system of detailed balance equations (3.35) is closed by the physical equations (3.9) and (3.19), which connect the macroscopic characteristics of plastic deformation with the characteristics of the dislocation structure. The initial conditions of the type $t = 0$, $\rho_g = \rho_{g0}$, where ρ_{g0} is the initial density of mobile dislocations in the material are recorded. Overcoming of the barriers by the dislocations during plastic deformation is a process that is thermally and mechanically activated. Therefore, for frequencies v_{sg} the equation (3.14) is valid. The solution of the indicated system of equations is the scalar law of plastic flow.

The authors of the approach do not concretize the mechanisms of dislocation transformations in their works, therefore they only qualitatively describe and analyze the regularities of the plastic flow of various materials. For the quantitative description of the deformation diagrams of a particular material, the frequencies of dislocation transformations are selected from the condition that the experimental and theoretical dependences coincide [161]. The approach also uses the averaged characteristics of the dislocation structure, therefore, in fact, it is phenomenological.

In concluding this section, we note the following. In the physics of strength and plasticity, the task is not to develop models of plastic deformation, suitable for solving applied problems related to the calculation of stresses and deformations, for example, for calculating and mathematical modelling of technological processes of plastic forming of metals (forging, stamping, pressing, rolling, drawing, etc.). In the physics of strength and plasticity, there is no method of generalizing the uniaxial flow law to a triaxial one. The main task is the theoretical study of deformation as a physical process. Therefore, the developed models of high scientific importance, for example, given in [69], can not be used in engineering practice because of their cumbersomeness, complexity, the presence of parameters whose physical meaning remains unclear or does not have sufficiently accurate estimates.

The construction of engineering models on their basis is the task of specialists in applied sciences. In this sense, the above approach by B.A. Grinberg and M.A. Ivanov to the description of plastic deformation seems, in our opinion, very promising because of, first of all, simplicity and clarity and, secondly, due to the maximum proximity to the kinetic method. As already noted, plastic deformation and destruction is a single irreversible process. Therefore, its adequate theoretical description is possible only with the use of methods of kinetics.

A complete description of the plastic deformation must contain the flow law and the model of ultimate deformation. This thesis is true and for constructing the theory of creep of metals.

3.2.4. Physical (microstructural) models of creep of metals

Microstructural models of creep of metals are considered in monographs [44, 45]. They give a wide panorama of the results of creep research, including phenomenology and physical mechanisms, mathematical description and analysis of the physical meaning of the parameters of creep equations. The effect on the creep rate of temperature, applied stress and material structure is analyzed.

The phenomenology of creep was considered in 1.4.2. Here, focusing on papers [44, 45], we confine ourselves to only some microstructural models of creep that will be of interest when considering the physical and mathematical theory of creep of metallic materials.

The main modern microstructural models of creep are based on the assumption that creep, as a kind of irreversible deformation, is caused by slip of dislocations. In the case of creep of pure metals, various versions of the Weertman model were widely used [44].

They are based on the assumption that the rate of stationary creep is controlled by the creep of dislocations with the predominant role of their slip. When gliding, dislocations stop at barriers and form accumulations. The climb of the leading dislocations and their annihilation control the creep rate, for which, at the steady state, the following expression was derived[1]

$$\dot{\varepsilon}^c = A \frac{D_L}{b^{3.5} M_u^{0.5}} \left(\frac{\sigma^c}{G} \right)^{4,5} \frac{Gb^3}{kT}, \qquad (3.36)$$

where A is a dimensionless constant; D_L is the volume diffusion coefficient; M_u is the density of Frank–Read sources.

---

[1] All the below microstructural creep models describe the creep rate at the steady state.

The model (3.36) has been criticized. In particular, it was claimed that in dislocation clusters were never observed in the samples after creep.

Assuming that not individual dislocations creep, but their complexes climb, Weertman obtained the equation

$$\dot{\varepsilon}^c = A_c \frac{D_L}{b^2} \left(\frac{\sigma^c}{G}\right)^3 \frac{G\Omega}{kT}, \qquad (3.37)$$

where $\Omega = 0.75b^3$ is the atomic volume.

The dependence $\dot{\varepsilon}^c \sim \sigma^3$ is considered «natural» in dislocation creep models, in which it is assumed that the diffusion flows go in the volume of grains.

Climb of dislocations is a recovery mechanism. It always requires diffusion. Assuming that the diffusion takes place both over the volume of grains (volume diffusion) and along the dislocation cores, the equation for the creep rate at the steady state (minimum velocity) takes the form

$$\dot{\varepsilon}^c = A \left[D_L + B \left(\frac{\sigma^c}{G}\right)^2 D_c \right] \left(\frac{\sigma^c}{G}\right)^3 \frac{G\Omega}{b^2 kT}, \qquad (3.38)$$

where A and B are constants, $A = 11$, $B = 3, 0$; D_c is the diffusion coefficient for the dislocation nuclei.

The described models belong to the group of dislocation creep models, in which $\dot{\varepsilon}^c$ is controlled by a diffusion-dependent recovery. Their generalization leads to an equation that is valid in the case of bulk diffusion:

$$\dot{\varepsilon}^c = A_L \frac{D_L}{b^2} \left(\frac{\sigma^c}{G}\right)^3 \frac{G\Omega}{kT}, \qquad (3.39)$$

and to equation

$$\dot{\varepsilon}^c = A_c \frac{D_L}{b^2} \left(\frac{\sigma^c}{G}\right)^5 \frac{G\Omega}{kT}, \qquad (3.40)$$

for the case of diffusion along the nuclei of dislocations [44].

An essential shortcoming of these models is the impossibility of describing unsteady creep processes in cases of plane and volumetric stress states.

A separate type of creep is *diffusion creep*, at which the mass transfer is carried out by individual atoms, i.e., as a result of diffusion in the stress field. In this case, diffusion of vacancies and a counterflow of atoms can occur both in the volume of grains and along their boundaries. In the latter case, the activation energy is much smaller.

The physical nature of diffusion creep as a mechanism of irreversible deformation according to modern concepts is as follows. At grain boundaries, under the influence of tensile stresses σ, the energy of vacancy formation is reduced by $\sigma\Omega$, where Ω is the atomic volume. At the boundaries experiencing compressive stress, it is increased by the same amount. As a result, a gradient of vacancy concentration appears in the grain volume, which is the driving force of their movement. A motion of vacancies, which is directed in accordance with the direction of the concentration gradient, and, consequently, the counterflow of atoms (Fig. 3.25) form. This process leads to a change in the size and shape of the grain, that is, to its plastic deformation. In the zone of compressed boundaries, the crystallite is 'disassembled' and 'built up' in the zone of stretched boundaries (shaded in Fig. 3.25).

There are several types of diffusion (vacancy) creep, including Nabarro–Herring creep, when diffusion flows along the body of grains (as depicted in the diagram), and Cobble creep with diffusion flows along grain boundaries.

The effect of this or that variety of DPs upon deformation of a particular material is determined, as always, by the amount of energy consumed by the mechanism. The energy of diffusion creep is mainly related to the value of the diffusion coefficient and the magnitude of the diffusion path. In ultrafine-grained materials, the Cobble mechanism predominantly operates.

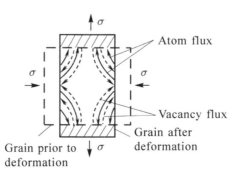

Fig. 3.25. The atomic mechanism of diffusion creep.

The phenomenological equation of diffusion creep corresponds to the equation of a Newton ideal-viscosity liquid

$$\sigma^c = \mu \dot{\varepsilon}^c, \qquad (3.41)$$

where μ has the meaning of the viscosity coefficient.

The microstructural (physical) approach, which is based on model concepts of the atomic deformation mechanism, makes it possible to determine the phenomenological coefficient μ in (3.41) through the characteristics of the metal structure.

Physical equations of diffusion creep:

a) Nabarro–Herring creep, when volume diffusion predominates,

$$\sigma^c = \frac{d^2 kT}{B\Omega D_L} \dot{\varepsilon}^c \qquad (3.42)$$

or

$$\dot{\varepsilon}^c = \frac{B\Omega D_L \sigma^n}{d^2 kT}, \qquad (3.43)$$

where B is a constant;

b) Cobble creep, when grain boundary diffusion prevails, is described by Eq. (3.24) or

$$\dot{\varepsilon}^c = \frac{B_c w \Omega \sigma^c D_{gb}}{d^3 kT}, \qquad (3.44)$$

where B_c is a constant; D_{gb} is the grain-boundary diffusion coefficient.

Diffusion creep, when $\dot{\varepsilon}^c$ is proportional to the first degree σ, is observed for very small σ and in practice is very rare. Therefore, equations (3.42)–(3.44) have basically only theoretical value.

It has been experimentally established that creep at high homological temperatures and low stresses has the features of diffusion creep (the dependence $\dot{\varepsilon}^c \sim \sigma^c$) and at the same time $\dot{\varepsilon}^c$ does not depend on the grain size d. This creep was called the Harper–Dorn creep [44] (by the first name of the authors).

On the basis of the results of the study of this creep, the following conclusions were drawn: a) dislocations participate in deformation, and their density does not depend on σ^c; b) a substructure is formed in the creep process; c) $\dot{\varepsilon}^c$ is almost three orders of magnitude higher than the rate predicted by the Nabarro–Herring equation (3.43).

Barrett and co-authors proposed a creep equation derived under the following assumptions: a) the multiplication of dislocations occurs by climb, and the deformation is the result of slip of dislocations; b) the slip velocity of the dislocations $v_g \sim u_0 \sigma^c \sim D_L \sigma^c$; c) the average subgrain size $d \sim G/\sigma^c$; d) the density of dislocation sources M_s is independent of σ^c; e) annihilation of dislocations takes place at the boundaries of subgrains.

In this case, the following equation is obtained for the dislocation density

$$\rho = M_s \frac{d_0}{v_0} \frac{D_L b}{GkT}, \qquad (3.45)$$

where $d_0 = d\sigma^c$ is a constant [44].

If we assume that the creep rate is described by the Orowan equation (3.19), then we obtain

$$\dot{\varepsilon}^c = M_s d_0 \frac{D_L b^2}{GkT} \sigma . \qquad (3.46)$$

The Harper–Dorn creep is very rare. It is believed that the creep rate of pure metals at a temperature $T > 0.5\ T_m$ is controlled by bulk diffusion. In the case of a solid solution consisting of components A and B, the situation becomes more complicated, since here, along with the mutual diffusion coefficient D, one must consider the partial diffusion coefficients of both components, D_A and D_B, and the self-diffusion coefficients of these components D_A^* and D_B^*.

If the steady-state creep rate $\dot{\varepsilon}^c \sim \sigma^3$ and there is no dependence on the stacking fault energy and the inversed primary creep stage, these solid solutions belong to class I. For solid solutions of class II, $\dot{\varepsilon}^c \sim \sigma^5$, the strong dependence of $\dot{\varepsilon}^c$ on the stacking fault energy packaging and a pronounced stage of normal primary creep (stage of unsteady creep) are typical [44].

The creep properties of solid solutions are explained by the interaction of dislocations with impurity atoms. There are four interaction mechanisms [70]: 1) the change in the energy of the stacking faults; 2) elastic interaction (Cottrell interaction); 3) chemical interaction (Suzuki effect); 4) interaction of dislocations with short-range ordering (Fisher's interaction).

Impurity atoms slow the movement of dislocations, forming around them impurity atmospheres. In this case, the dislocations entrain the impurity atoms, which makes their motion viscous.

As an example, we give an equation for the rate of steady-state creep, controlled by the viscous motion of dislocations, which was proposed by Weertman [44]:

$$\dot{\varepsilon}^c = \frac{2\pi(1-v)G}{A}\left(\frac{\sigma}{G}\right)^3, \qquad (3.47)$$

where v is Poisson's ratio; A is a constant that depends on the temperature and mechanism of interaction of dislocations with impurity atoms.

It turns out that the creep of solid solutions, as well as of pure metals, is most often characterized by the dependence $\dot{\varepsilon}^c \sim \sigma^c$. However, the creep rate of the solid solution, other things being equal, is lower than the creep rate of the base pure metal.

The development of technology requires the development of structural materials that are characterized by high values of creep resistance, that is, they are able to maintain high strength at elevated operating temperatures. In this respect, solid-solution hardening is ineffective.

The main modern direction of development of heat-resistant materials is the creation of *dispersed* and *dispersion-hardened* systems. In these materials, the creep resistance is ensured by the presence of a second finely divided solid phase in them. There are two technologies for their creation. The dispersed phase is introduced into the melt of the matrix mechanically or by sintering powders and represents foreign particles. This is *dispersion hardening*. These alloys satisfy the definition of *composite materials*.

In superalloys the dispersed phase is separated from the solid solution with appropriate heat treatment. This is *dispersion hardening*. Heat-resistant nickel-based alloys hardened by intermetallic particles are widely used. The size of the dispersion particles is from several hundreds to thousands of angstroms (1 Å = 10^{-8} cm), and the average distance between them is of the order of a tenth of a micron. These particles are effective barriers for dislocations. No movement of dislocations – no irreversible deformation.

In superalloys, high creep resistance is provided by solid-solution hardening (the effect of dissolved elements) and, to a large extent, by the presence of a second dispersed phase.

To date, several authors have proposed several microstructural creep models for dispersed and dispersion-hardened alloys, which are discussed in monographs [44, 45]. The models are based on various ideas about the interaction of moving dislocations with dispersed particles and mechanisms for their overcoming.

The main features of the creep of alloys with dispersed phases are as follows [45]:

1) the behaviour of alloys is practically independent of the way in which dispersed phases are obtained in it;
2) under other identical conditions, the creep rate of alloys with dispersed phases is much less than that of one matrix;
3) the exponent n in the $\dot{\varepsilon}^c \sim \sigma^n$ dependence is always greater than that of the pure metal, and is usually 7–8;
4) the activation energy of creep is almost 2 times higher than the activation energy of self-diffusion in the matrix.

It has been established theoretically and experimentally that the particle size and density are the main influence on creep and, accordingly, heat resistance. The general regularity is as follows: the bulk density of dispersed particles is smaller and their size is less than $\dot{\varepsilon}^c$ and higher the heat resistance. There is no generally accepted theory that has good agreement with experiment.

In conclusion, we present some information on the regularities of the formation and evolution of a dislocation structure under creep. A common regularity is the formation of a substructure at the beginning of the established stage. Numerous results of experimental studies, generalized in [44], show that the dependence of the scalar density of free dislocations on the stress is described by the formula

$$\rho = B'\left(\frac{\sigma}{Gb}\right)^p, \qquad (3.48)$$

where B' is a constant. The value of exponent p is in the range {1–2}. It is proved that the density of dislocations does not change during the steady-state creep. It is believed that this fact determines the stage of stationary creep with $\dot{\varepsilon}^c_{min}$.

For the linear size of subgrains, an empirical relationship is obtained, which is also valid for active deformation:

$$D = B/\sqrt{\rho}, \qquad (3.49)$$

where B is a constant, in order of magnitude equal to 10.0.

Combining (3.48) and (3.49) with $p = 2.0$, we obtain

$$D = B^*Gb/\sigma. \qquad (3.50)$$

Summing up the review of microstructural (physical) creep models of metals, which is not claimed for completeness, it is necessary to note the following. Creep is a complex physical process, depending on a large set of both external thermomechanical factors, and on the initial structure and laws of its evolution. This explains the great variety of the proposed process models and their grouping by the type of metals and alloys.

The known models of creep are mainly of scientific importance, since they help to understand the physics of the process. The practical importance of models can not yet be recognized as significant, since the level of their development does not allow us to develop engineering methods for designing calculations of machine parts and structures for creep and, consequently, the long-term strength associated with it.

The process of creep, as well as the process of active irreversible deformation, is accompanied by accumulation in the metal of discontinuities (deformation damage) and ends with destruction.

3.3. Basic concepts and provisions of the physics of fracture of metals

Following the sequence of stages objectively existing in the scientific method of cognition (collection of the results of observations, classification and description, phenomenology of the phenomenon, physical concepts of the phenomenon), we will consider the types of macroscopic destruction observed in the technology. At the same time, we shall rely mainly on fundamental work [72].

The simplest of them is plastic slipping (in our opinion, the more successful term is plastic shear) along the plane of easy sliding (Fig. 3.26 *a*). It is observed when single crystals are destroyed. In polycrystals, destruction by shear is more common (Fig. 3.26 *b*). The rupture of the metal into two parts occurs as a result of the formation of a large number of microcracks along the localized flow plane with subsequent destruction of the bridges between them. It is identified by the characteristic surface of the fracture on which a system of elongated ellipsoidal pits is observed.

The most dangerous in technology is brittle fracture – the growth of a brittle crack from a stress concentrator or a macroscopic defect of the notch type (Fig. 3.26 *c*). The plastic deformation is very small. The fracture surfaces with a purely brittle character consist of a set of atomically smooth facets with a crystallographic orientation during transcrystalline fracture or sections of intergranular boundaries under intergranular fracture.

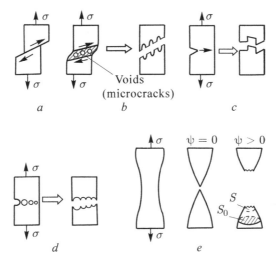

Fig. 3.26. Types of macroscopic fracture: *a* – shift; *b* – shear; *c* – growth of a brittle crack; *d* – growth of a viscous crack; *e* – the loss of macroscopic stability [72].

Destruction, accompanied by large plastic strains ($\varepsilon^p > 0.1$ viscous destruction) can occur in two forms: fracture from a ductile crack (Fig. 3.26 *d*) and failure due to loss of macroscopic stability (deformation localization) (Fig. 3.26 *e*).

With an intermediate quasi-brittle failure, when a plastic zone forms near the crack tip, the failure surfaces correspond to the plastic case and represent a chaotic relief with a large number of protrusions and valleys.

The main concept in the physics of the destruction of solids is the concept of an *embryonic microcrack* – an elementary carrier of body damage. A microcrack is understood to be a discontinuity with a characteristic linear dimension $h < h_{Gr} = 2\gamma\sigma E/\sigma^2$ (see (2.3)). Therefore, the nucleation and growth of a microcrack in a field of homogeneous stresses is energetically disadvantageous [72]. Microcracks can appear only due to stress concentrators, primarily due to inhomogeneous plastic deformation and the work of external forces on this deformation.

In plastic materials, which are metal alloys subjected to pressure treatment, local shear stresses near the tip of the microcrack cause plastic deformation. As a result, so-called plastic zones appear that move ahead of the top of the crack when it is opened and the formation of which requires additional energy.

In this case, in continuation of the investigation of the phenomenology of destruction begun in Section 2.1, the growth condition of the microcrack will look like this:

$$h > h_{\text{Or}} = [2(\gamma + \gamma_{\text{pl}})E/\sigma^2],\qquad(3.51)$$

where γ_{pl} is the specific work of plastic deformation in the plastic zone per unit surface of the microcrack; h_{Or} is the critical size according to Orowan.

In this case $h_{\text{Or}} \cong (10^2 - 10^4) h_{\text{Gr}}$. Therefore, it can be assumed that the plastic materials are destroyed not as a result of the growth of a single microcrack to the size of the main crack h_σ, for which a breakdown criterion is satisfied ($\sigma_{\text{loc}} \geq \sigma_{\text{theor}}$) and which already spontaneously propagates in the material at the speed of sound, but due to the growth of the number of microcracks and their subsequent fusion in the macrocrack ($h_{\text{Or}} \leq h < h_\sigma$), for which the destruction energy condition is fulfilled, but the force condition is not yet met, or directly into the main crack ($h > h_\sigma$).

Before we briefly dwell on modern ideas about microscopic mechanisms of destruction, we formulate, following [72], the main points arising from the consideration of the phenomenology of the phenomenon, set out in particular in section 2.1.

1. The process of destruction is multi-stage. The transition from one stage to another occurs when the crack reaches some critical lengths determined by the structure of the metal and the balance of stresses and energies. At each stage of the process, the intrinsic microscopic mechanism of the main mechanism.

2. The nucleation and growth of cracks obey the power and energy criteria. When the force criterion is satisfied, the interatomic bonds are torn apart in a time of the order of the period of atomic oscillations $t \sim v_D^{-1}$. This condition is sufficient. The fulfillment of the energy criterion means only the possibility of a break, therefore it is necessary. At intermediate stresses $\sigma_w < \sigma_{\text{loc}} < \sigma_{\text{theor}}$ only the thermal activation mechanism of destruction is possible, i.e. in the general case the destruction is a kinetic process.

3. The processes of destruction and plastic deformation at the level of micromechanisms are interrelated. Plastic deformation promotes the initiation of microcracks with $h < h_{\text{Gr}}$ and inhibits the growth of cracks with $h > h_{\text{Gr}}$. Therefore, *a single physical process of plastic deformation and ductile fracture of metals should be considered.*

4. In metals, the character of the forces of interatomic bonds is collective. Groups of 20–40 atoms interact among themselves. Therefore, the elementary act of the process of fracture must be collective, involving at the same time several tens of neighbouring atoms.

These provisions are the foundation for the study, analysis and theoretical description of a single process of plastic deformation and the destruction of metals. At the same time, looking ahead, we note that the requirement of a single theoretical model becomes, in this case, simultaneously one of the criteria for its adequacy.

To initiate microcracks, gradients of strains and stresses are necessary and, consequently, an inhomogeneous distribution of the elastic energy stored in the material. The natural mechanism for creating these gradients during plastic deformation, as noted in section 3.1, is the mechanism of dislocation sliding and the formation of dislocation clusters in individual microvolumes. Then the sequence of nucleation of microcracks appears in the following form: the accumulation of external forces in the deformed volume of work in the form of elastic energy of dislocations; transfer of this energy in volume by dislocation flows; its concentration in the places of maximum gradients of plastic deformation; the relaxation of energy by the creation of new free surfaces, i.e., by the initiation of microcracks.

To date, several mechanisms for the relaxation of elastic energy associated with dislocations have been developed and quantitatively described (by means of the formation of new surfaces, i.e. dislocation mechanisms of microcrack nucleation). They are considered in detail in many monographs on the problem of ductile fracture [72, 73].

Let us consider the model of the mechanism of thermal fluctuation initiation of a microcrack in the head of a retarded planar dislocation cluster, developed by the author [72]. In our opinion, this mechanism is dominant in various metals and alloys, since a dislocation cluster can form on various barriers. This mechanism will be taken in the future as the basis for the construction of a unified physico-mathematical theory of deformation and ductile fracture of metals.

The author [72] states that the idea of the possibility of microcrack initiation in the head of the inhibited shear belongs to Zener, and the first mathematical model was developed by Stroh. The model is based on the property of a flat dislocation cluster (Fig. 3.12) – the presence of high local stresses in its head (3.15).

To bring two ($n = 2$) similar edge dislocations closer to a distance $d \sim 10b$, in accordance with (3.17), for example, at $G = 80\,000$ MPa, a stress of $\tau \sim 1274$ MPa is required, which is several times larger than that usually observed with hot and warm deformation and is the maximum permissible at cold deformation. But with the accumulation of $n \geq 20$ dislocations, the approach of the leading dislocations to a distance of $\sim 10b$ becomes real at ordinary stresses. Dislocations of the same name, located at a distance $d > (3-4)b$, repel, so to merge their nuclei it is necessary to overcome the potential barrier. Its height is proportional to the length of the interacting dislocations. Consequently, the combination initially can occur on a short section of dislocation lines.

The process of microcrack nucleation appears to consist of several stages (Fig. 3.27). Plane clusters of $n = 20-30$ or more dislocations are observed experimentally in metals. Therefore, the approach of the leading dislocations to a distance $d = (3-4)b$ is a real event (Fig. 3.27 a). Under conditions of warm or hot deformation, a thermal fluctuation ejection of a short $(2-3)b$ region of the second dislocation is possible, on which relaxation relaxation begins – the merger of the nuclei (Fig. 3.27 b).

At the second stage, a short microcrack is generated as a result of combining the nuclei of two dislocations which then expands, as shown in Fig. 3.27 c, d. The remaining dislocations of the cluster are then transferred into the nucleated microcrack which then grows to the value $h \cong n^2 b/2$ (Fig. 3.27 d).

In this model, the elementary act of nucleation of a microcrack is the ejection of a pairwise kink on the second dislocation, the merger of the nuclei of the pairwise kink and the corresponding portion of the leading dislocation, the expansion of a short microcrack and its growth to $h \cong n^2 b/2$ due to the transfer of the remaining dislocations of the cluster into it. The activation energy of this elementary act is determined mainly by the activation energy of the formation of a pairwise kink. Theoretical estimates of the activation energy, carried out in [186], gave the value of $U_0 \cong Gb^2/2$, i.e., approximately equal to the activation energy of sublimation.

In the case of plastic deformation, U decreases due to stress work σ and is equal to

$$U(\sigma) = U_0 - n\sigma V, \qquad (3.52)$$

where n acquires the meaning of the overstress factor; $V = (5-8)b^3$ is the activation volume.

Fundamentals of the Physics of Strength and Plasticity   133

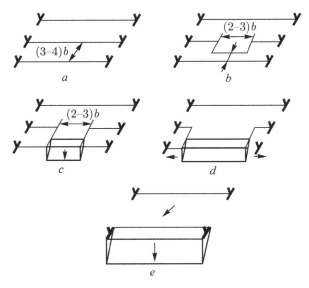

Fig. 3.27. Scheme of successive stages of thermal fluctuation initiation of a microcrack in the head of an inhibited cluster of dislocations [72].

The condition for the thermal fluctuation nucleation of the microcrack acquires the form

$$n\tau \sim \sigma_{theor} = G/2\pi, \tag{3.53}$$

where σ_{theor} is the theoretical tensile strength.

In this case, the size of the emerging microcrack will be equal to

$$h \cong n^2 b/2 \cong 3G^2 b/8(\pi\sigma)^2. \tag{3.54}$$

The ratio $h/h_{Gr} \cong 3 \cdot G^2 b \cdot 4\sigma^2 / 8\pi^2 \sigma^2 \cdot 2.6 G^2 b \cong 0.1$, that is, $h \cong 0.1 h_{Gr}$, and the microcrack is stable.

The estimate of h for (3.54) and for the intermediate condition (3.53), $n\tau \cong 0.5 G/2\pi$, for example, for low-carbon steels at hot deformation temperatures ($G \approx 40000$ MPa and $\sigma = 120$ MPa) gives $h \cong 10^{-5}$ cm. Detailed experimental studies of the statistics of microcracks in viscous fracture in broken samples showed that the average size of microcracks is 10^{-5} cm [187–189]. These authors proposed to call these microcracks *submicrocracks*.

Since the nucleation of dispersed mechanically stable submicrocracks with $h < h_{Gr}$, the process of destruction has just begun. The development of the process is determined by the thermomechanical conditions of deformation and the main parameters of the microcracks. Depending on these conditions, the microcracks

134 Physico-Mathematical Theory of High Irreversible Strains

can grow or heal, passing into other lattice defects. One of the fundamental provisions of the physical theory of fracture is the provision that the microcrack, along with vacancies, dislocations, interstitial atoms, is one of the elements of the defective structure of the crystal lattice. *The fundamental property of defects is their mutual transformability.*

Let us consider qualitatively, following [72], possible mechanisms for healing the microcrack, that is, transforming it into more elementary defects of the lattice.

The main parameters of the dislocation microcrack are its Burgers vector $\mathbf{B} = n\mathbf{b}$ and the size (length) h (Fig. 3.28). The potential energy is determined by the value of B. Therefore, the relaxation processes leading to a decrease in the potential energy of the microcrack are associated with a change in B (decrease in B).

Figure 3.29 shows possible mechanisms for the transformation of a dislocation microcrack into other lattice defects. After the removal of the applied stress, it is possible to emit dislocations near the base – a process reverse to the 'dumping' of the dislocations in the submicrocrack nucleated in the head of a flat cluster (Fig. 3.29 *a*). When the dislocations are emitted near the top (Fig. 3.29, *b*), the microcrack is blunted, transformed into a pore, and mechanically stabilized.

The microcrack can turn into a dislocation wall due to the climbing of dislocations at its base (Fig. 3.29 *c*). This mechanism is long and requires high temperatures, which ensure a high concentration of equilibrium vacancies. The dislocation microcrack can be transformed into a void due to absorption of vacancies by the tip (Fig. 3.29 *d*). The most probable process is the transformation of the microcrack into a microvoid due to surface diffusion. The flow of vacancies is

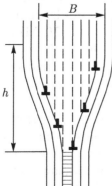

Fig. 3.28. The representation of a dislocation crack in the form of the distribution of splitting dislocations [72].

Fundamentals of the Physics of Strength and Plasticity

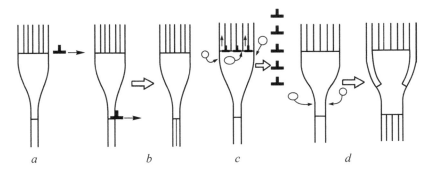

Fig. 3.29. Mechanisms for the transformation of a dislocation microcrack into other lattice defects: *a* – the emission of dislocations near the base; *b*) emission of dislocations near the tip; *c*) spreading of dislocations of the microcrack and its transformation into a dislocation wall; *d* – the absorption of vacancies by the tip [72].

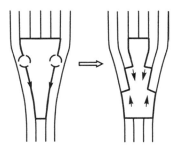

Fig. 3.30. The transformation of a dislocation microcrack into a void due to surface diffusion [72],

directed from the base to the top, and the flow of atoms is from the top to the bottom (Fig. 3.30). The emergence of vacancies on the surface and their migration along it require an activation energy that is about half that of analogous processes in the bulk of the crystal.

Since the dislocation microcrack is a stress concentrator, it quickly blunts and evolves by one of the mechanisms considered in the micropores. The voids are more stable than the dislocation microcracks. However, surplus surface energy is also associated with them. Evolution of the voids, like microcracks, is a thermally activation process, therefore is largely determined by the thermomechanical conditions of deformation. Possible mechanisms for the transformation of microvoids, including other lattice defects, are shown in Fig. 3.31.

First of all, the voids tends to assume a spherical shape due to surface diffusion (Fig. 3.31 *a*). The void can migrate as a whole

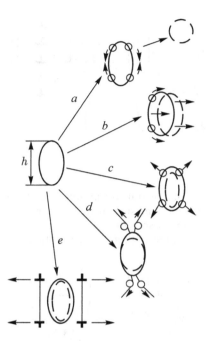

Fig. 3.31. Mechanisms of void evolution: *a* – change in shape under the influence of surface diffusion; *b* – displacement of the void as a whole (creep) due to surface diffusion; *c* – dissolution by the emission of vacancies and their diffusion into the volume of the crystal: *d* – dissolution by the emission of vacancies along dislocations and their diffusion along dislocation tubes; *r* – void collapse due to the emission of prismatic dislocation loops of interstitial type under the action of compressive stresses (hydrostatic pressure) [72].

due to the origin of vacancies on one side of it and the emergence of their directed flow to the other due to the presence, for example, of a stress gradient.

The disappearance of voids due to dissolution by the emission of vacancies is of the greatest practical interest.

The driving force of dissolution is surface tension. Here two options are possible: the emission of vacancies into the surrounding volume (Fig. 3.31 *c*); pipe diffusion of vacancies along dislocation cores (Fig. 3.31 *d*). If the void was formed as a result of the evolution of a dislocation microcrack, then, as a rule, it is associated with a dislocation system.

Void dissolution processes require a long time and a high temperature. Therefore, they are important in the annealing of metals and sintering. In the processes of plastic deformation, the mechanism of void healing by emitting prismatic dislocation loops under the action of a positive hydrostatic pressure $p = -\sigma_0$, where σ_0 is the stress of the spherical component of the stress tensor (Fig. 3.31, d) is apparently the most real and essential. In the future, we will consider this mechanism to be dominant in the healing of deformation microscopic discontinuities during the process of plastic deformation of metals.

In connection with the concept of the multilevel nature of plastic deformation, the authors of Ref. [190] attempted a theoretical analysis of the conditions for the formation of a microcrack in highly deformed crystals based on the formation of fragmented structures due to plastic rotations caused by the motion of partial disclinations.

The information given above from the field of strength and plasticity physics is sufficient for constructing and comprehending a unified physico-mathematical theory of plasticity and deformability of metals.

Section II

New Physico-Mathematical Theory of High Irreverseible Strains and Ductile Fracture of Metals

4

A physico-phenomenological model of the single process of plastic deformation and ductile fracture of metals

4.1. General provisions of the model

The structural–phenomenological approach allowed to formulate the fundamental principles that led to the construction of a unified theory of irreversible deformation, ductile fracture and evolution of the microstructure [191]. The theory is based on the following generally accepted assumptions of the physics of strength and plasticity, which were analyzed in Chapter 3.

1. The dominant deformation mechanism in a sufficiently wide range of variation in the strain rate $\dot{\varepsilon}$ and temperature T is the gliding of dislocations in the grains. The range includes temperatures and rates at which the metal forming processes are realized in the industry.

2. The resistance to deformation is caused by the braking of the flow of mobile dislocations by barriers on which they stop and become stationary. The main barriers are dislocation type barriers: stationary dislocations (forest dislocations), subgrain, grain and interphase boundaries. The cause of hardening is an increase in the deformation of the density of stationary dislocations.

3. Softening is associated with a decrease in the density of the stationary dislocations due to these dislocation overcoming some of the barriers and their transformation again into mobile ones (contributing to deformation), by annihilation during climbing around barriers by creep or double cross slip and encountering a dislocation of the opposite sign, and also by the formation of new grain boundaries during recrystallization and microdisruption of continuity (deformation damage) by dislocation mechanisms. At the same time, the necessary energy for overcoming the barriers (activation energy) is obtained by the stationary dislocations through thermal activation and the work of acting stresses.

4. The softening micromechanisms – the climbing of dislocations (non-conservative motion) and the ejection of double kinks in the local section of the dislocation line to another crystallographic plane for cross slip have a diffusion nature, hence the height of the potential barrier is equal to the activation energy of self-diffusion. The activation energy for the nucleation of the submicrocrack is close to the sublimation energy of the metal [72].

5. Irreversible deformation and ductile fracture is a single kinetic multistage process in which the nucleation of diffuse nucleation submicrocracks with an average size of $\overline{\xi} = 10^{-5}$ cm occurs as a dislocation mechanism with the onset of deformation. Submicrocracks are generated in most cases in the head of a retarded planar dislocation cluster, i.e., the complex of immobile (stationary) dislocations becomes a nucleation submicrocrack.

6. The deformation damage of the metal can be reduced due to the disappearance (healing) of microcracks by the predominantly sequential operation of two micromechanisms: the transformation of the dislocation microcrack into microvoids due to surface diffusion; collapse of the void due to the emission of prismatic dislocation loops of interstitial type under the action of compressive stresses (hydrostatic pressure, Fig. 3.31). These micromechanisms are thermally and mechanically activated, have a diffusion nature and the activation energy, as noted, is equal to the activation energy of self-diffusion.

7. When the scalar density of microcracks of critical value is reached, their integration into a macrocrack takes place – a crack visible to the naked eye, and this act is interpreted as the fracture of the metal.

The main principles of the physics of the unified process of irreversible deformation and ductile fracture of metals are based on the fundamental property of crystal lattice defects (point, linear, planar and volume) – their mutual transformation.

Let us generalize these generally accepted provisions in the form of two postulates.

Postulate 1. *Irreversible deformation and ductile fracture of metals is a single kinetic and multistage process that takes place under conditions of competitive processes of hardening and softening, nucleation and disappearance of dislocation nucleation microcracks – elementary structural carriers of deformation damage.*

The content of the formulated propositions and the postulate clearly illustrates the scheme of mutual transformations of defects of the crystal lattice under irreversible deformation, shown in Fig. 4.1.

When the yield stress reaches the yield point for the given thermomechanical loading conditions, the dislocation sources begin to generate mobile dislocations g, the directed motion of which in the stress field carries out mass transfer and, accordingly, plastic deformation.

After sliding through the path from the source to the barrier (the mean free path), the mobile dislocations g stop at them and become stationary dislocations s. This dislocation transformation occurs with the frequency v_{gs} averaged over the volume. An increase in the scalar density of the stationary dislocations ρ_s takes place and is the main reason for the hardening of the metal under deformation. The decrease in the density of stationary dislocations (softening) can occur in three directions (Fig. 4.1). They can overcome barriers and turn again into mobile ones with a frequency v_{sg}, annihilate bypassing

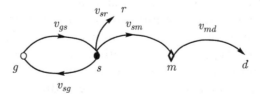

Fig. 4.1. Scheme of mutual transformation of crystal lattice defects in a single process of irreversible deformation and ductile fracture of metals.

barriers and spend to form new grain boundaries with a frequency v_{sr}, and also to turn into submicrocracks with a frequency v_{sm}.

The emergence of dispersed submicrocracks is accompanied by the start of the first stage of fracture, beginning almost simultaneously with the appearance of plastic flow [68, 72]. Under certain thermomechanical conditions of deformation the microcracks can disappear (heal). This act occurs with the frequency v_{md} averaged over the volume (Fig. 4.1). With increasing deformation (in the second stage of fracture), the density of microcracks increases with the speed determined by the ratio of the frequencies v_{sm} and v_{md}. The third stage of fracture occurs when the density of microcracks reaches a critical value $N_m = N^* = 10^{11} \ldots 10^{12}$ cm^{-3} [187]. The average size of nucleation microcracks is $\bar{\xi} = 10^{-5}$ cm [187], therefore, their density can be determined, by analogy with the scalar density of dislocations, as the total length of all microcracks per unit volume. Then $N^* = 10^6 \ldots 10^7$ cm^{-2}. At this density, according to the existing ideas, the microcracks are combined into a macrocrack.

Note that the scheme shown in Fig. 4.1 is approximate. For a consistent description of the kinetics of accumulation and transformation of crystal lattice defects, it is necessary to consider in detail the interaction of ensembles of defects of various types with the details of the atomic mechanisms of their mutual transformation. However, as will be shown below, taking into account the most essential features of a single process of plastic deformation and fracture, it allows us to obtain simple equations that ensure the solution of practical problems with practice accuracy.

The above scheme makes it easy to give a mathematical formulation of postulate 1 in the form of kinetic balance equations for the dislocation and microcrack densities, the evolution equations for the actual structural parameters of the metal, which are the average scalar density of stationary dislocations ρ_s and microcracks N_m. For the steady-state deformation, the equations have the form

$$\frac{d\rho_s}{dt} = \rho_g v_{gs} - \rho_s v_{so}, \tag{4.1}$$

$$\frac{dN_m}{dt} = \xi_0 \rho_s v_{sm} - N_m v_{md}, \tag{4.2}$$

where $\xi_0 = 10^{-5}$ is a dimensionless quantity taking into account the different geometric form of dislocations and microcracks and numerically equal to the average length of microcracks $\bar{\xi}$; t is time.

Since the activation energies of self-diffusion and sublimation of metals are very close and the concrete stationary dislocation s disappears by one of three possible transformations ($s \to g$, $s \to r$, $s \to m$), then for the sake of simplifying equation (4.1), which is the basis for further reasoning, by v_{so} we mean the mean frequency of dislocation disappearance s, which is determined by the activation energy of self-diffusion.

The ratio of hardening (the first term on the right side of equation (4.1)) and softening (the second term) determines the value of the flow stress σ and depends on the thermomechanical conditions of deformation. Therefore, in accordance with the above provisions 3 and 4, we formulate the second assertion.

Postulate 2. *For the non-equilibrium process of irreversible deformation of metals, the relationship between stress σ, strain rate $\dot{\varepsilon}$ and temperature T is determined by thermomechanically activated micromechanisms of hardening and softening and is described by the equation*

$$\dot{\varepsilon} = \dot{\varepsilon}_0 \exp\left(-\frac{U - \tau V}{kT}\right). \tag{4.3}$$

In this equation, $\dot{\varepsilon}_0$ is a factor that has the meaning of the rate of plastic deformation of a crystal by the mechanism of slip of dislocations in the ideal case of the absence of barriers; U is the activation energy of a micromechanism controlling the rate of deformation.

In accordance with the contents of the above provisions 1–4: $U = \beta G b^3$ – activation energy of self-diffusion [68]; $\beta = 0.38$–0.48 is the empirical coefficient depending on temperature, and for some alloys – on deformation (for carbon steels and some alloys $\beta(T) = 5 \cdot 10^{-5} T + 0.3639$); G is the shear modulus; b is the average modulus of the Burgers vector of dislocations (interatomic distance, atomic size (in estimates for metals, $b = (2-3) \cdot 10^{-8}$ cm), τV is the work of acting stresses, decreasing the height of the potential barrier U, $\tau = \sigma/m$ is the intensity of the shear stress, σ is the stress intensity, m is the Taylor factor for polycrystals (for chaotic misorientation of the grains $m = 3.1$), $V = b^2/\sqrt{\rho_s}$ is the activation volume (the volume of the dislocation segment is a monatomic chain of atoms of length $1/\sqrt{\rho_s}$), which must be activated in order to increase the probability of overcoming the barried by the climb of the segment, double cross sliding or the formation of a nucleation submicrocrack; k is the Boltzmann constant.

For the first time, as noted in 3.1, the position on the effect of thermal vibrations of atoms on the characteristics of the strength of metals was formulated by Becker [146, 147]. To account for the non-equilibrium of the deformation process, the second term, the specific work of the acting stresses, is introduced into the exponential (4.3).

Equation (4.3) reduces to the form

$$\sigma = \left[\beta Gbm - \frac{kTm}{b^2} \ln \frac{\dot{\varepsilon}_* b \sqrt{\rho_s}}{\dot{\varepsilon}} \right] \sqrt{\rho_s}, \qquad (4.4)$$

where $\dot{\varepsilon}_* b \sqrt{\rho_s} = \dot{\varepsilon}_0$; $\dot{\varepsilon}_*$ is numerically equal to the Debye frequency $v_D \cong 10$ s^{-1} [67] (the frequency of thermal vibrations of the ion in the crystal lattice practically constant for all metals).

To obtain a closed system of equations, we add four equations to (4.1), (4.2) and (4.4), which, like (4.4), describe the macroscopic characteristics of deformation and fracture with the characteristics of the structure known in the physics of strength and plasticity (see (3.14) and (3.19)) and are included in the group of equations that form the basis of this discipline:

$$\dot{\varepsilon} = \rho_g bv, \qquad (4.5)$$

$$v_{so} = v_0 \exp\left(-\frac{U - \sigma V_{so}/m}{kT} \right), \qquad (4.6)$$

$$v_{sm} = v_{sm}^0 \exp\left(-\frac{U_{sm} - \sigma V_{sm}/m}{kT} \right), \qquad (4.7)$$

$$v_{md} = v_{md}^0 \exp\left(-\frac{U_{md} - pV_{md}/M}{kT} \right), \qquad (4.8)$$

where U, U_{sm}, U_{md} are activation energies of micromechanisms of softening, nucleation and healing of microcracks, respectively; $U = U_{md}$ is the activation energy of self-diffusion; $U_{sm} = \beta_s Gb^3$, where $\beta_s = 0.48$–0.58 is the energy of sublimation; V_{so}, V_{sm}, V_{md} are the corresponding activation volumes; $p = -\sigma_0$ is hydrostatic pressure; $\sigma_0 = \sigma_{ii}/3$ is the average normal stress; M is a factor that takes into account the different crystallographic orientation of microcracks (in the case of chaotic misorientation, one can expect $M = m = 3.1$); v is the average slip velocity of dislocations in the presence of barriers.

The meaning of the Orowan equation (4.5) is simple – the rate of plastic deformation is proportional to the flow of mobile dislocations $\rho_g v$. The exponents in the right-hand sides of equations (4.6)–(4.8) have the meaning of the probabilities of disruption of a stationary dislocation from the barrier, the formation of a nucleation microcrack and the healing of a microcrack, respectively. Temperature and stress-independent pre-exponential factors v_0, v_{sm}^0, v_{md}^0 are the number of attempts per unit time taken by a stationary dislocation to pass the barrier, ejection of a double kink by a second dislocation in an inhibited flat cluster to initiate a microcrack, and a prismatic dislocation loop from the micropores when it is healed, respectively. Here $v_0 = v_{sm}^0$ and is equal to the frequency of the thermal oscillations of the dislocation segment of length $1/\sqrt{\rho_s}$, in which there are $1/\sqrt{\rho_s}\, b$ ions. One ion in the lattice oscillates with a frequency v_D, and the oscillation frequency of $1/\sqrt{\rho_s}\, b$ ions is $v_D b \sqrt{\rho_s}$.

The value of v_{md}^0 is estimated on the basis of the following considerations. In the case of a disk-shaped microvoids v_{md}^0 is the frequency of thermal oscillations of a monatomic closed (in the form of a circle) chain of ions of a crystal lattice of diameter $\bar{\xi}$ and a length of $\pi\bar{\xi}$, in which there are $\pi\bar{\xi}/b$ ions. One ion in the lattice oscillates with a frequency v_D, and the frequency of thermal vibrations of $\pi\bar{\xi}/b$ ions is $v_{md}^0 = v_D b / \pi\bar{\xi}$.

The activation volumes $V_{so} = V_{sm} = b^2/\sqrt{\rho_s}$ and the activation volume of the prismatic dislocation loop leaving the microvoids of diameter $\bar{\xi}$ under hydrostatic pressure $V_{md} = \pi\bar{\xi} b^2$.

By the solution of the system of equations (4.1)–(4.8), taking into account the fact that $v/v_{gs} = \lambda$ is the mean free path of dislocations and $dt = d\varepsilon/\dot{\varepsilon}$, the scalar functionals of the flow stress and microcrack density,

$$\sigma = \left[\beta(T)G(T)bm - \frac{kT(\varepsilon)m}{b^2} \ln \frac{\dot{\varepsilon}_* b \sqrt{\rho_s(\varepsilon)}}{\dot{\varepsilon}(\varepsilon)} \right] \times$$

$$\times \left\{ \int \left[\frac{1}{b\lambda} - \frac{(\rho_s(\varepsilon))^{3/2} v_D b}{\dot{\varepsilon}(\varepsilon)} \cdot \exp \right. \right. \quad (4.9)$$

$$\left. \left. \left(-\frac{\beta(T)G(T)b^3 - \sigma b^2/m\sqrt{\rho_s(\varepsilon)}}{kT(\varepsilon)} \right) \right] d\varepsilon \right\}^{1/2},$$

Physico-Phenomenological Model of Single Process

$$N_m = \int \left[\xi_0 \rho_s(\varepsilon) \cdot \frac{v_D b \sqrt{\rho_s(\varepsilon)}}{\dot{\varepsilon}(\varepsilon)} \exp \right.$$
$$\left(-\frac{\beta_s(T)G(T)b^3 - \sigma b^2/m\sqrt{\rho_s(\varepsilon)}}{kT(\varepsilon)} \right) - \quad (4.10)$$
$$\left. -N_m \cdot \frac{v_D b}{\pi \bar{\xi} \dot{\varepsilon}(\varepsilon)} \exp\left(-\frac{\beta(T)G(T)b^3 + \sigma_0 \pi \bar{\xi} b^2/M}{kT(\varepsilon)} \right) \right] d\varepsilon.$$

4.2. The scalar defining equation of viscoplasticity

Irreversible deformation and ductile fracture of metals from a thermodynamic point of view are a single non-equilibrium process. Consequently, σ and N_m depend not only on the values of T and $\dot{\varepsilon}$ at the considered finite time, but also on the law of their change at previous instants of time, that is, on $T(\varepsilon)$ and $\dot{\varepsilon}(\varepsilon)$. Therefore, for a numerical calculation of the functionals (4.9) and (4.10) and taking into account the loading history, we replace them with a system of equations in finite increments.

To this end, we differentiate (4.4) with respect to ρ_s on a small interval $d\varepsilon$, assuming on it T, $\dot{\varepsilon}$ = const:

$$d\sigma_{(g)} = \left[\frac{\beta m G b}{2\sqrt{\rho_{s(g)}}} - \frac{mkT_{(g)}}{2b^2 \sqrt{\rho_{s(g)}}} \left(1 + \ln \frac{\dot{\varepsilon}_* b \sqrt{\rho_{s(g)}}}{\dot{\varepsilon}_{(g)}} \right) \right] d\rho_{s(g)}. \quad (4.11)$$

The index (g) here and below will denote the number of the calculated step g = 1, 2, ..., n with step-by-step calculation of the deformation diagram $\sigma(\varepsilon)$ of the material. In this case, at each step g, the deformation intensity will obtain a small finite increment of the quantity $d\varepsilon(g)$ = 0.001–0.01. At each calculation step, $\dot{\varepsilon}$ and T can take different values, but within the limits of the step, because of the smallness of $d\varepsilon$, they are assumed to be constant.

Equation (4.1) can be written in the form

$$d\rho_{s(g)} = \left[\frac{1}{b\lambda} - \frac{\rho_{s(g-1)}^{3/2} v_D b}{\dot{\varepsilon}_{(g)}} \exp\left(-\frac{\beta G b^3 - \sigma_{(g-1)} b^2/m\sqrt{\rho_{s0}}}{kT_{(g)}} \right) \right] d\varepsilon_{(g)}. \quad (4.12)$$

The meaning of equation (4.12) is as follows. The first term

$d\varepsilon_{(g)}/b\lambda$ is the increment of the density of the stationary dislocations at an arbitrary calculated step g with the deformation increment by the amount $d\varepsilon_{(g)}$ in the absence of any softening (recovery) processes. The second term is the number of stationary dislocations that disappeared at step g due to the work of the above three softening mechanisms.

We note that the use of equations in increments for the formulation of the theory of irreversible deformation does not presently complicate, but simplifies the solution of practical problems. The initial-boundary problems of plasticity and creep, including three-dimensional ones, are currently being solved using computer software developed on the basis of the numerical finite element method. The method uses a step-by-step calculation algorithm. Therefore, the model in increments and the numerical method of finite elements are organically consistent with each other.

On the basis of (4.4), we write the equation for determining the initial (initial) yield point for specific values of $\dot{\varepsilon}$ and T:

$$\sigma_{(1)}^T = \left[\beta m G b - \frac{k T_{(1)} m}{b^2} \ln \frac{\dot{\varepsilon}_* b \sqrt{\rho_{s0}}}{\dot{\varepsilon}_{(1)}} \right] \sqrt{\rho_{s0}}, \qquad (4.13)$$

where ρ_{s0} is the initial (before heating and deformation) dislocation density in the material.

The first term in (4.13) is the yield strength of the metal under cold deformation. At the next calculated step, $g = 2$, the strain rate and T can change, so we write the equation for determining the yield point at an arbitrary step g (more precisely, the initial flow stress at step g) as

$$\sigma_{(g)}^T = \left[\beta m G b - \frac{k T_{(g)} m}{b^2} \ln \frac{\dot{\varepsilon}_* b \sqrt{\rho_{s(g-1)}}}{\dot{\varepsilon}_{(g)}} \right] \sqrt{\rho_{s(g-1)}}. \qquad (4.14)$$

This stress also determines the value of the elastic deformation at the beginning of the calculated step g as $\varepsilon_{(g)}^e = \sigma_{(g)}^T / E$.

From equations (4.11), (4.12) and (4.14) it follows that at an arbitrary step g, characterized by the increment $d\varepsilon(g)$, the flow stress is represented in the form

$$\sigma_{(g)} = \sigma_{(g)}^T + d\sigma_{(g)}. \qquad (4.15)$$

For a step-by-step calculation of the deformation diagram $\sigma(\varepsilon)$ the equations (4.14), (4.12), (4.11) and (4.15) are supplemented by the obvious relations

$$\rho_{s(g)} = \rho_{s(g-1)} + d\rho_{s(g)}, \tag{4.16}$$

$$\varepsilon_{(g)} = \varepsilon_{(g-1)} + d\varepsilon_{(g)}. \tag{4.17}$$

In this case, $\sigma_{(g)}$ found by (4.15) is put in correspondence with $\varepsilon_{(g)}$, determined from (4.17).

The substitution of (4.12) into (4.11) yields

$$d\sigma_{(g)} = d\sigma^u_{(g)} - d\sigma^r_{(g)} \text{ and } d\sigma^u_{(g)} = \frac{\beta m G b}{2\sqrt{\rho_{s(g)}}b\lambda} d\varepsilon_{(g)}, \tag{4.18}$$

where $d\sigma^u_{(g)}$ is the athermal component of $d\sigma_{(g)}$ due to hardening;

$$d\sigma^r_{(g)} = \left\{ \frac{\beta m G b^2 \rho^2_{s(g)} v_D}{2\dot{\varepsilon}_{(g)}} \exp\left(-\frac{\beta G b^3 - \sigma_{(g-1)}b^2/m\sqrt{\rho_{s0}}}{kT_{(g)}}\right) + \right.$$

$$+ \frac{mkT_{(g)}}{2b^3\lambda\sqrt{\rho_{s(g)}}} \left(1 + \ln\frac{\dot{\varepsilon}_*b\sqrt{\rho_{s(g)}}}{\dot{\varepsilon}_{(g)}}\right) \cdot \left[1 - \frac{\rho^2_{s(g)}b^2 v_D\lambda\sqrt{\rho_{s(g)}}}{\dot{\varepsilon}_{(g)}} \times \right. \tag{4.19}$$

$$\times \exp\left(-\frac{\beta G b^3 - \sigma_{(g-1)}b^2/m\sqrt{\rho_{s0}}}{kT_{(g)}}\right)\right]\right\} d\varepsilon_{(g)}$$

is the thermal component $d\sigma_{(g)}$, due to the processes of softening. The last component determines the dependence of stress on $\dot{\varepsilon}$ and T, i.e., describes the viscosity of the material.

We write (4.15) in the form

$$\sigma_{(g)} = \sigma^T_{(g)} + d\sigma^u_{(g)} - d\sigma^r_{(g)}. \tag{4.20}$$

It follows from (4.20) that the scalar defining equation of a viscoplastic medium, which has the form of an operator in increments (4.14), (4.16), (4.17), (4.18), (4.19) and (4.20), based on two postulates, describes a wide spectrum of the rheological behaviour of metals, determined by the ratio of the three components of the stress. This ratio depends on the nature of the metal and the thermo-speed deformation conditions. For example, for $d\sigma^u_{(g)} = d\sigma^r_{(g)}$ in (4.20) and $\sigma^T_{(g)} = \text{const}$, we have an ideal plastic medium; at $d\sigma^r_{(g)} = 0$ the model (4.20) describes the hardening body; for $d\sigma^r_{(g)} > d\sigma^u_{(g)}$ we have a non-monotonic diagram $\sigma(\varepsilon)$ with a falling section. The model also describes the $\sigma(\varepsilon)$ dependences for various laws of variation of $\dot\varepsilon$ and T in the deformation process.

4.3. Scalar model of the plasticity of a hardening body (cold deformation of metals)

By way of illustration of the foregoing, let us consider the model of a hardening rigid-plastic body (cold metal deformation), which is obtained as a special case of the model of viscoplasticity described. In the case of cold deformation, dislocations overcome barriers by the force method [192–194], in which the entire activation energy $U = \beta Gb^3$ in (4.6) is ensured by the operation of acting stresses,

$$\beta Gb^3 = \sigma b^2 / m\sqrt{\rho_s}.$$

From this equality it follows, firstly, that for the cold deformation equation (4.4) takes the form

$$\sigma = \beta mGb\sqrt{\rho_s}. \tag{4.21}$$

Secondly, in the case of force overcoming of barriers, the probability of overcoming them in (4.6) is 1.0 (exp 0 = 1). In this case, as follows from (4.3), $\dot\varepsilon = \dot\varepsilon_0 = \sqrt{\rho_s}v_D b = \sqrt{\rho_s}\dot\varepsilon_* b$ and (4.12) takes the form

$$d\rho_s = \left[\frac{1}{b\lambda} - \rho_s\right]d\varepsilon.$$

Integrating this equation under the initial conditions $\varepsilon = 0$, $\rho_s = \rho_{s0}$, we obtain

$$\rho_s = \frac{(b\lambda)^{-1}[\exp(\varepsilon)-1]+\rho_{s0}}{\exp(\varepsilon)}. \tag{4.22}$$

The substitution of (4.22) into (4.21) gives the scalar defining equation of the hardening rigid-plastic body in the form of a finite functional dependence [195]

$$\sigma = \beta mGb\left\{\frac{(b\lambda)^{-1}[\exp(\varepsilon)-1]+\rho_{s0}}{\exp(\varepsilon)}\right\}^{1/2}. \tag{4.23}$$

The initial yield stress for a rigid-plastic body at $\varepsilon = 0$ will be equal to

Physico-Phenomenological Model of Single Process 149

$$\sigma_T = \beta m G b \sqrt{\rho_{s0}}. \qquad (4.24)$$

The model of cold deformation allows us to propose a basic experiment for determining two parameters of the general model ρ_{s0} and λ necessary for its practical use.

The values of ρ_{s0} and λ are characteristics of the initial dislocation structure of metals. Therefore, they can be determined by the methods of microstructural analysis. However, for mechanics it is simpler to construct a true metal deformation diagram according to the results of standard tests by sedimentation of cylindrical specimens with grooves at the ends under cold deformation conditions. With its use, the sought-for characteristics are calculated by formulas

$$\rho_{s0} = \frac{\left(\sigma_T^{\exp}\right)^2}{(\beta m G b)^2}, \quad \lambda = \frac{b(\beta m G)^2 [\exp(\varepsilon)-1]}{\sigma^2 \exp(\varepsilon) - (\beta m G b)^2 \rho_{s0}},$$

where $\sigma_{T\exp}^{}$ is the experimentally determined yield point; ε and σ are is the strain intensity from the interval (0.1–0.5) and the stress intensity value corresponding to it in the experimental diagram. These formulas are obtained from (4.23) and (4.24).

The physico-mathematical theory of a large cold cyclic and close to it deformation is presented in Chapter 5.

4.4. Model of ductile fracture of metals

For a stepwise calculation of the scalar density of microcracks arising during deformation (deformation damage), the functional (4.10) is also replaced by a system of equations in finite increments:

$$dN_{m(g)} = \left[\xi_0 \rho_{s(g)} \frac{v_D b \sqrt{\rho_{s(g)}}}{\dot{\varepsilon}_{(g)}} \exp\left(-\frac{\beta_s G b^3 - \sigma_{(g)} b^2 / m \sqrt{\rho_{s(g)}}}{kT}\right) - \right.$$
$$\left. - N_{m(g-1)} \frac{v_D b}{\pi \bar{\xi} \bar{\varepsilon}_{(g)}} \exp\left(-\frac{\beta G b^3 + \sigma_{(g)} K_{(g)} \pi \bar{\xi} b^2 / mM}{kT}\right) \right] d\varepsilon_{(g)}, \qquad (4.25)$$

$$N_{m(g)} = N_{m(g-1)} + dN_{m(g)}, \qquad (4.26)$$

where $K(g) = \sigma_{0(g)}/\tau_{(g)}$ is the rigidity index of the stress state [2] (characterizes the fraction of stresses of the spherical tensor in the stressed state), uniaxial tension and compression $K = 0.58$ and -0.58, respectively, with shear $K = 0$).

If the stressed state of the deformed body at the calculated step g is characterized primarily by compressive stresses, that is, $\sigma_{0(g)} < 0$, then $p_{(g)} > 0$, $K_{(g)} < 0$ and the activation energy of microvoid healing $U_{md} = \beta G b^3$ in (4.25) is reduced by the value of the work $\sigma_{(g)} K_{(g)} \pi \bar{\xi} b^2 / mM$, performed by the hydrostatic pressure $p_{(g)}$. If tensile stress prevails in the stress state, that is, $\sigma_{0(g)} > 0$, then $K_{(g)} > 0$ and the activation energy of healing of the microvoids in (4.25) increases, and the probability of its healing, respectively, decreases.

Since N_m is a scalar, the physical conditions for the deformation of the elementary volume at the calculated step g without fracture and with fracture will be written accordingly in the form

$$N_{m(g)} < N^*_{(g)}, N_{m(g)} = N^*_{(g)}.$$

The extent of use of a part of the plasticity resource for g calculated steps on which it accumulated the strain rate $\varepsilon_{(g)}$, or the probability of macrofracture, can be defined as $\psi_{(g)} = N_{m(g)}/N^*_{(g)}$. Then the conditions of deformation of an elementary volume without fracture and with fracture take the form

$$\psi_{(g)} = N_{m(g)}/N^*_{(g)} < 1.0, \qquad (4.27)$$

$$\psi_{(g)} = N_{m(g)}/N^*_{(g)} = 1.0. \qquad (4.28)$$

The equations of the ductile fracture model (4.25)–(4.28) include the characteristics of plastic deformation, which indicates the unity of these processes.

The density operator of microcracks (4.25)–(4.26) describes the ductile fracture, like the plasticity operator, over a wide range of variation of T and $\dot{\varepsilon}$. Let us consider a special case of the theory, which is of great practical importance – ductile fracture during cold deformation [196, 197]. Cold stamping of metals occupies an important place in the technology of pressure metal working and in the processes of plastic structure formation.

It is known that for cold deformation ($T < 0.2 T_m$, where T_m is the melting point), the flow stress and plasticity (deformation before fracture) are practically independent of the strain rate $\dot{\varepsilon}$ (in the interval of variation $\dot{\varepsilon}$ in which dynamic effects can be neglected) The strain rate can be eliminated from equation (4.25), if we put

Physico-Phenomenological Model of Single Process 151

$$v_{sm(g)} = v_D b\sqrt{\rho_{s(g)}} \exp\left(-\frac{\beta_s G b^3 - \sigma_{(g)} b^2/m\sqrt{\rho_{s(g)}}}{kT_{(g)}}\right) = k_{sm(g)}\dot{\varepsilon}_{(g)}, \quad (4.29)$$

$$v_{md(g)} = \frac{v_D b}{\pi\bar{\xi}} \exp\left(-\frac{\beta G b^3 + \sigma_{(g)} K_{(g)} \pi\bar{\xi} b^2/mM}{kT_{(g)}}\right) = k_{md(g)}\dot{\varepsilon}_{(g)}, \quad (4.30)$$

where $k_{sm(g)}$ and $k_{md(g)}$ are proportionality coefficients.

An analysis of the dimensions of the quantities in the right and left sides of equations (4.29) and (4.30) shows that

$$v_D b\sqrt{\rho_{s(g)}} = \dot{\varepsilon}_{(g)}, \; v_D b/\pi\bar{\xi} = \dot{\varepsilon}_{(g)}, \quad (4.31)$$

$$k_{sm(g)} = \exp\left(-\frac{\beta_s G b^3 - \sigma_{(g)} b^2/m\sqrt{\rho_{s(g)}}}{kT_{(g)}}\right), \quad (4.32)$$

$$k_{md(g)} = \exp\left(-\frac{\beta G b^3 + \sigma_{(g)} K_{(g)} \pi\bar{\xi} b^2/mM}{kT_{(g)}}\right). \quad (4.33)$$

The following statements follow from these equations.

1. The work of the mechanisms of nucleation and healing of microcracks in the cold deformation conditions at $\dot{\varepsilon}_{(g)} = (10^{-2}-10^3)$ s^{-1}, which are realized in materials testing and cold stamping in industry, is equivalent to the operation of mechanisms under high temperatures ($T > 0.6T_m$) with the rates $\dot{\varepsilon} = v_D b\sqrt{\rho_{s(g)}} \cong v_D b/\pi\bar{\xi} \cong 10^{10}$ s^{-1} [1]). This is due to the fact that at $\dot{\varepsilon} \cong 10^{10}$ s^{-1} the processes of return and healing of microcracks do not have time to flow and deformation at high T is realized by the mechanism of cold deformation.

2. Equations (4.31) show the dependence of the average size of the nucleated microcracks on the dislocation density of the form

$$\bar{\xi} \cong 1/\pi\sqrt{\rho_s}. \quad (4.34)$$

If, after metal annealing, the residual dislocation density $\rho_{min} = 10^8$ cm^{-2}, and after a large cold strain and $\rho_{max} = 10^{12}$ cm^{-2}, then the estimate from (4.34) gives $\bar{\xi}_{max} \cong 0.32 \cdot 10^{-4}$ cm and $\bar{\xi}_{min} \cong 0.32 \cdot 10^{-6}$ cm. It turns out that with increasing deformation and with increasing

---

[1]) The estimation was carried out at values of the quantities: $v_D = 10^{13}$ s^{-1}; $b = 3 \cdot 10^{-8}$ cm; $\rho_{s(g)} = 10^{10}$ cm^{-2}; $\pi = 3.14$; $\bar{\xi} = 10^{-5}$ cm.

152 *Physico-Mathematical Theory of High Irreversible Strains*

dislocation density, the average size of the emerging submicrocracks decreases and the theory developed correctly predicts the average size of the nucleated submicrocracks $\bar{\xi} = 10^{-5}$ cm, which is observed experimentally [187–188].

3. It follows from the above equations that the coefficients ksm and kmd are the probabilities of nucleation and healing of microcracks, respectively.

4. Conditions for the force nucleation and healing of microcracks follow from (4.32) and (4.33) in the form $k_{sm} = k_{md} = 1.0$ and

$$\beta_s G b^3 = \sigma_{(g)} b^2 / m \sqrt{\rho_{s(g)}},$$

$$\beta G b^3 = -\sigma_{(g)} K_{(g)} \pi \bar{\xi} b^2 / mM,$$

or

$$\sigma_{(g)} = \beta_s m G b \sqrt{\rho_{s(g)}}, \tag{4.35}$$

$$\frac{\sigma_0}{\tau} = K_{(g)} = -\beta G b m M / \sigma_{(g)} \pi \bar{\xi}. \tag{4.36}$$

Equation (4.35) is analogous to equation (4.21) and describes cold deformation. Consequently, when cold deformation occurs, when the force barriers overcome by the stationary dislocations, the force nucleation of dislocation submicrocracks takes place, and $k_{sm} = 1.0$. Physically, this means that in an delayed flat dislocation cluster, the number of dislocations n reaches a value when the intensity of tangential stresses in the head of the cluster reaches the value of the theoretical shear strength.

5) It follows from (4.36) that for values of the stiffness index of the stressed state $K^* = \sigma_0^*/\tau = -\rho^*/\tau \leqslant -\beta G b m M / \sigma \pi \bar{\xi}$, metals should exhibit practically unlimited plasticity, since all microcracks should be healed by high compressive stresses. An estimate of the K^* value for a low-carbon steel in which $\beta = 0.38$, $G = 84800$ MPa, $b = 3 \times 10^{-8}$ cm, $m = M = 3.1$, $\sigma = 300$ MPa, shows that K^* essentially depends on the value of $\bar{\xi}$. For $\bar{\xi} = 10^{-5}$ cm (mean experimental value), $K^* \leq -0.992$; at $\bar{\xi} = 0.32 \cdot 10^{-5}$ cm (the average theoretical value, determined by the formula (4.34)) $K^* \leq -3.1$. On the experimental plasticity diagrams $\Lambda^*(K)$, where Λ^* is the limiting (up to fracture) shear strain, the values of K^* at which the tendency of the asymptotic aspiration of Λ^* to infinity is visible, for aluminium alloys are -0.992, for low-carbon steels -3.1 [2, 104]. Therefore, for steels we take $\bar{\xi} = 0.32 \cdot 10^{-5}$ cm. We note that a relation of the type (4.36) was first obtained in [72].

Substituting equations (4.31)–(4.33) into (4.25), we obtain for the cold deformation

$$dN_{m(g)} = \left[\xi_0 \rho_{s(g)} - N_{m(g-1)} \cdot \exp\left(-\frac{\beta G b^3 + \sigma_{(g)} K_{(g)} \pi \bar{\xi} b^2 / mM}{kT}\right)\right] d\varepsilon_{(g)}. \quad (4.37)$$

The first term in (4.37) $\xi_{0\rho s(g)} \, d\varepsilon_{(g)}$ is the increment in the number of microcracks originating at step g. The second term is the number of microcracks healed in step g by the mechanism described above.

The expression for the healing probability of microcracks (4.33) can be simplified by substituting equations (4.34) and (4.35) into it. After substituting and transforming, we get

$$k_{md(g)} = \exp\left(-\beta G b^3 (1 + K_{(g)}/M)/kT_{(g)}\right). \quad (4.38)$$

Taking into account (4.38), the equation for the increment of the scalar density of microcracks at the loading step g, characterized by the increment of the strain intensity $d\varepsilon_{(g)}$, will take the final form

$$dN_{m(g)} = \left\{\xi_0 \rho_{s(g)} - N_{m(g-1)} \cdot \exp\left[-\beta G b^3 (1 + \frac{K_{(g)}}{M})/kT_{(g)}\right]\right\} d\varepsilon_{(g)}, \quad (4.39)$$

where ξ_0 is a dimensionless quantity numerically equal to $\bar{\xi}$ = 0.32 · 10⁻⁵.

Equations (4.39), (4.26), (4.27) and (4.28) are the model of ductile fracture of metals under cold deformation, which is an integral part of the unified plasticity and ductile fracture model, since $\rho_{s(g)}$ and $K_{(g)}$ in (4.39) are calculated within the plasticity model.

In contrast to the well-known phenomenological criteria of authors' damage: *Cockroft–Latham, Freudenthal, Rice–Tracy, Oyane, Ayada, Osakada, Brozzo, Zhao–Kuhn*, which are incorporated into the DEFORM program complex of computer modeling of plastic deformation processes [123], the proposed theory takes into account the processes of nucleation and healing of microcracks and the influence of load history on the accumulation of microcracks due to the dependences $\rho_{s(g)}(t)$, $K_{(g)}(t)$ and $T_{(g)}(t)$.

4.5. Obtaining a generalized law of viscoplasticity based on a scalar law

To apply the above theory in solving practical problems, for example, for calculating and mathematical modeling of the stress–strain state and predicting the probability of workpiece destruction during the development of the process of metal working with pressure, it is necessary to have a tensor defining equation – the generalized flow law. The generalization was carried out using the method adopted in the classical mathematical theory of plasticity [8, 59].

From the latter, the postulates of the existence in stress space of the loading function and the von Mises plasticity conditions are borrowed. The difference is that in the general case of a viscoplastic body, the loading function is not a limiting surface, the introduction of which in the mathematical theory of plasticity is consistent with the concept of a complete deformation in the form of a sum of elastic and plastic strains, $\varepsilon_{ij} = \varepsilon_{ij}^e + \varepsilon_{ij}^p$, but is represented as an instantaneous surface in a six-dimensional space the stress tensor separating at each instant of time (at the calculated step g) regions with higher stresses $\sigma_{ij(g)}$ outside the surface (from the surface) from the region (inside the surface) lower $\sigma_{ij(g)}$. In the special case of an elastoplastic hardening material, the instantaneous loading surface becomes limiting and separates the region of elastic states from the plasticity region.

Instantaneous at the loading step g, the loading function, in accordance with the physical scalar defining relation (4.20), is taken in the form of the instantaneous von Mises plasticity condition in the form

$$f_{(g)} = f_{(g)}^T + df_{(g)}^u - df_{(g)}^r = \frac{3}{2}\left[s_{ij(g)}^T + ds_{ij(g)}^u - ds_{ij(g)}^r\right] \times$$
$$\times \left[s_{ij(g)}^T + ds_{ij(g)}^u - ds_{ij(g)}^r\right] = \left[\sigma_{(g)}^T + d\sigma_{(g)}^u - d\sigma_{(g)}^r\right]^2, \quad (4.40)$$

where $s_{ij(g)}^T$, $ds_{ij(g)}^u$ and $ds_{ij(g)}^r$ are the deviators of the tensors $\sigma_{ij(g)}^T$, $d\sigma_{ij(g)}^u$ and $d\sigma_{ij(g)}^r$, respectively; $\sigma_{(g)}^T$, $d\sigma_{(g)}^u$ and $d\sigma_{(g)}^r$ are the stress intensities of the corresponding stressed states, determined at each loading step g by formulas (4.14), (4.18) and (4.19), respectively.

Along with the processes of active loading, unloading and neutral loading known in the mathematical theory of plasticity for a viscoplastic medium, the concept of the thermodynamic return process is introduced. Consider its geometric interpretation, using the six-dimensional space representing the tensor of the acting stress, as is customary in the mathematical theory of plasticity.

We shall consider the deformation of the element of the medium at temperatures and strain rates at which in (4.20) $d\sigma^r_{(g)} \neq 0$ and, accordingly, $d\sigma^r_{ij(g)} \neq 0$, that is, the deformation of the viscoplastic body. For the body, as follows from (4.14) and (4.19), the unloading process has no physical meaning. If the stress is reduced during deformation, the particle will not pass from the plastic state to the elastic state, and the deformation will continue at a smaller value of $\dot{\varepsilon}$.

Assume that at the beginning of the loading step g, as follows from (4.20), the stress state of the particle is described by the vector $\bar{\sigma}^T_{(g)}$, whose end, in accordance with (4.14), lies on the instantaneous loading surface $\Sigma^T_{(g)}$ (Fig. 4.2). When the viscoplastic body is deformed, we can single out the process of softening due to the component $d\sigma^r_{(g)}$ in (4.20), and the active loading process associated with the component $d\sigma^u_{(g)}$, which is characteristic of a hardening plastic body. The set of these processes is shown in Fig. 4.2.

Since the work of the softening micromechanisms is due to thermomechanical activation, the process of voltage reduction associated with $d\sigma^r_{ij(g)}$ is called a thermomechanical return. In this process, the component $d\bar{\sigma}^r_{(g)}$ of the docking vector $d\bar{\sigma}_{(g)}$ is directed inside the instantaneous loading surface, but the vector $\bar{\sigma}^T_{(g)} - d\bar{\sigma}^r_{(g)}$ remains on the loading surface, which is compressed, i.e., the plasticity condition is satisfied.

When the viscoplastic body is deformed at the loading step g, the thermodynamic return and loading processes proceed simultaneously and, in accordance with (4.18) and (4.19), lead to an increase in the plastic deformation intensity $d\varepsilon_{(g)}$ and, consequently, to the

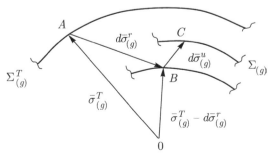

Fig. 4.2. Geometric interpretation of the thermodynamic return process in the stress space.

deformation tensor increment $d\varepsilon_{ij(g)}$. If the deformation occurs under conditions of time variations $\dot{\varepsilon}$ and T, then the loading surface pulsates, dividing the outer load region with high $\dot{\varepsilon}$ and low T from the inner load region with lower $\dot{\varepsilon}$ and high T.

The construction of the physico-mathematical theory of plasticity on the basis of the extreme principles of the classical mathematical theory of plasticity is impossible, since the thermodynamic return process contradicts the postulate and the stability condition of Drucker's deformation, and the von Mises and Ziegler maximum principles do not take into account the loading history [8].

Therefore, in order to construct a tensor law of deformation of a viscoplastic body within the framework of the flow theory, two theorems are formulated and proved.

Theorem 1. *The work of additional stresses on the increments of irreversible deformations of the viscoplastic body caused by them during the loading cycle and the thermodynamic recovery at the loading step g is positive:*

$$\left(\sigma^T_{ij(g)} - d\sigma^r_{ij(g)}\right) d\varepsilon_{ij(g)} + d\sigma^u_{ij(g)} d\varepsilon_{ij(g)} > 0. \qquad (4.41)$$

Proof of the lemma. The scalar model of viscoplasticity has the form

$$\sigma_{(g)} = \sigma^T_{(g)} + d\sigma^u_{(g)} - d\sigma^r_{(g)}.$$

Based on postulates 1 and 2, we have

$$\sigma^T_{(g)} = \beta m G b \left[1 - \frac{kT_{(g)} m}{\beta G b^3} \ln \frac{\dot{\varepsilon}_* b \sqrt{\rho_{s(g-1)}}}{\dot{\varepsilon}_{(g)}} \right] \sqrt{\rho_{s(g-1)}} > 0,$$

$$d\sigma^u_{(g)} = \frac{\beta m G b}{2\sqrt{\rho_{s(g)}} b \lambda} d\varepsilon_{(g)} > 0.$$

It can be argued that

$$\sigma^T_{(g)} + d\sigma^u_{(g)} > d\sigma^r_{(g)}.$$

Otherwise, if

$$\sigma^T_{(g)} + d\sigma^u_{(g)} = d\sigma^r_{(g)},$$

then $\sigma_{(g)} = 0$ and $d\varepsilon_{(g)} = 0$, that is, there is no deformation. Then

$$\left(\sigma^T_{(g)} - d\sigma^r_{(g)}\right) d\varepsilon_{(g)} + d\sigma^u_{(g)} d\varepsilon_{(g)} > 0.$$

Physico-Phenomenological Model of Single Process

Summarizing this scalar expression, we obtain

$$\left(\sigma^{T}_{ij(g)} - d\sigma^{r}_{ij(g)}\right)d\varepsilon_{ij(g)} + d\sigma^{u}_{ij(g)}d\varepsilon_{ij(g)} > 0. \tag{4.42}$$

If there are no thermodynamic returns, $d\sigma^{r}_{ij(g)} = 0$, then $\sigma^{T}_{ij(g)} = \sigma_{ij}$, $d\sigma^{u}_{ij(g)} = d\sigma_{ij}$, and (4.42) implies

$$\left(\sigma_{ij} + d\sigma_{ij}\right)d\varepsilon_{ij} > 0.$$

This is a well-known mathematical formulation of Drucker's postulate underlying the classical mathematical theory of plasticity. Consequently, theorem 1 is a generalization of the Drucker postulate on a viscoplastic medium.

Consequence. *The work of a part of the additional stresses $d\sigma^{u}_{ij(g)}$ which cause the process of loading when the viscoplastic body is deformed at the loading step g, is positive:*

$$d\sigma^{u}_{ij(g)}d\varepsilon_{ij(g)} > 0.$$

If we set $\sigma^{T}_{ij(g)} = d\sigma^{r}_{ij(g)}$ in (4.42), then we obtain $d\sigma^{u}_{ij(g)}d\varepsilon_{ij(g)} > 0$ or $d\sigma^{u}_{ij(g)}/d\varepsilon_{ij(g)} > 0$ is the deformation stability condition for the stress $\sigma^{u}_{ij(g)}$.

The established condition for the stability of deformation is also a generalization of the known Drucker stability condition for a viscoplastic medium.

Theorem 2. *For any given value of the components of the increment of irreversible deformation at the loading step g of a viscoplastic body, the increment of the work of irreversible deformation has a maximum value for the actual stress state determined by the load history ($\dot{\varepsilon}(t)$, $T(t)$) and the values $\dot{\varepsilon}_{(g)}$ and $T_{(g)}$, in comparison with all possible stress states with the same load history, allowed by the instantaneous loading function $f_{(g)}(\sigma_{ij(g)})$ and satisfying the condition*

$$f^{*}_{(g)}\left(\sigma^{T*}_{ij(g)}, d\sigma^{u*}_{ij(g)}, d\sigma^{r*}_{ij(g)}\right) < f_{(g)}\left(\sigma^{T}_{ij(g)}, d\sigma^{u}_{ij(g)}, d\sigma^{r}_{ij(g)}\right),$$

when $\dot{\varepsilon}^{*}_{(g)} < \dot{\varepsilon}_{(g)}$ and (or) $T^{*}_{(g)} > T_{(g)}$:

$$\left(\sigma^{T}_{ij(g)} - d\sigma^{r}_{ij(g)} + d\sigma^{u}_{ij(g)}\right)d\varepsilon_{ij(g)} > \left(\sigma^{T*}_{ij(g)} - d\sigma^{r*}_{ij(g)} + d\sigma^{u*}_{ij(g)}\right)d\varepsilon_{ij(g)}. \tag{4.43}$$

Proof of the lemma. On the basis of the postulates 1 and 2 for the stress intensities we have the relations

$$\sigma_{(g)}^{T*} < \sigma_{(g)}^{T}, d\sigma_{(g)}^{u*} = d\sigma_{(g)}^{u} \text{ and } d\sigma_{(g)}^{r*} > d\sigma_{(g)}^{r}$$

for $\dot{\varepsilon}_{(g)}^{*} < \dot{\varepsilon}_{(g)}$ and (or) $T_{(g)}^{*} > T_{(g)}$ (see equations (4.14), (4.18) and (4.19)). Therefore,

$$\left(\sigma_{(g)}^{T} - d\sigma_{(g)}^{r} + d\sigma_{(g)}^{u}\right)d\varepsilon_{(g)} > \left(\sigma_{(g)}^{T*} - d\sigma_{(g)}^{r*} + d\sigma_{(g)}^{u*}\right)d\varepsilon_{(g)}.$$

In this case, the non-negativity of the work of irreversible deformation (right and left of the inequality sign) is guaranteed by theorem 1.

Summarizing the last scalar inequality, we obtain

$$\left(\sigma_{ij(g)}^{T} - d\sigma_{ij(g)}^{r} + d\sigma_{ij(g)}^{u}\right)d\varepsilon_{ij(g)} > \left(\sigma_{ij(g)}^{T*} - d\sigma_{ij(g)}^{r*} + d\sigma_{ij(g)}^{u*}\right)d\varepsilon_{ij(g)}, \quad (4.44)$$

when $\dot{\varepsilon}_{(g)}^{*} < \dot{\varepsilon}_{(g)}$ and (or) $T_{(g)}^{*} > T_{(g)}$, which was to be proved.

If there are no thermodynamic return processes, $\sigma_{ij(g)}^{r} = 0$, then $\sigma_{ij(g)}^{T} + d\sigma_{ij(g)}^{u} = \sigma_{ij(g)}$ and from (4.44) we get: $\sigma_{ij}d\varepsilon_{ij} > \sigma_{ij}^{*}d\varepsilon_{ij}$, $\sigma_{ij}\dot{\varepsilon}_{ij} > \sigma_{ij}^{*}\dot{\varepsilon}_{ij}$ are the principles of the maximum work of plastic deformation and the rate of dissipation of mechanical work (the the von Mises principle), which underlie the classical mathematical theory of plasticity. Thus, theorem 2 is a generalization of the von Mises maximum principle to a viscoplastic medium.

The instantaneous loading function (4.40) and the generalized maximum principle (4.43) make it possible to write down the Lagrange associated flow law of a viscoplastic body at the loading step g in the form

$$\frac{\partial}{\partial\left(\sigma_{ij(g)}^{T} + d\sigma_{ij(g)}^{u} - d\sigma_{ij(g)}^{r}\right)}\left[\left(\sigma_{ij(g)}^{T} + d\sigma_{ij(g)}^{u} - d\sigma_{ij(g)}^{r}\right)d\varepsilon_{ij(g)} - d\lambda_{(g)}f_{(g)}\right] = 0, \quad (4.45)$$

$$d\varepsilon_{ij(g)} = d\lambda_{(g)}\partial f_{(g)}/\partial\left(\sigma_{ij(g)}^{T} + d\sigma_{ij(g)}^{u} - d\sigma_{ij(g)}^{r}\right),$$

where $d\lambda$ is the Lagrange multiplier.

It follows from (4.45) that $d\varepsilon_{ij(g)} = 3d\lambda_{(g)}\left(s_{ij(g)}^{T} + ds_{ij(g)}^{u} - ds_{ij(g)}^{r}\right)$, $d\varepsilon_{ii(g)} = 3d\varepsilon_{0(g)} = 0$, that is, the viscous plastic body satisfies the incompressibility condition. Therefore, the plastic deformation increment tensor is a deviator and

Physico-Phenomenological Model of Single Process 159

$$d\lambda_{(g)} = \frac{1}{2} \frac{d\varepsilon_{(g)}}{\sigma^T_{(g)} + d\sigma^u_{(g)} - d\sigma^r_{(g)}}.$$

The three-dimensional governing equations of a viscoplastic body instantaneous at the loading step g have the form

$$d\varepsilon_{ij(g)} = \frac{3}{2} \frac{d\varepsilon_{(g)}}{\sigma^T_{(g)} + d\sigma^u_{(g)} - d\sigma^r_{(g)}} \left(s^T_{ij(g)} + ds^u_{ij(g)} - ds^r_{ij(g)} \right). \qquad (4.46)$$

The principal difference between them and the equations of the mathematical theory of plasticity is that they are instantaneous – they describe an irreversible deformation at the step g, which is necessary to take into account the loading history.

To solve practical problems it is convenient to apply (4.46) in the form of three equations:

$$d\varepsilon_{ij(g)} = \frac{3}{2} \frac{d\varepsilon_{(g)}}{d\sigma^u_{(g)}} ds^u_{ij(g)}, \quad d\varepsilon_{ij(g)} = \frac{3}{2} \frac{d\varepsilon_{(g)}}{\sigma^T_{(g)}} s^T_{ij(g)},$$

$$d\varepsilon_{ij(g)} = \frac{3}{2} \frac{d\varepsilon_{(g)}}{d\sigma^r_{(g)}} ds^r_{ij(g)}.$$

This becomes clear if (4.46) is rewritten in the form

$$s^T_{ij(g)} + ds^u_{ij(g)} - ds^r_{ij(g)} = \frac{2}{3} \frac{d\varepsilon_{ij(g)}}{d\varepsilon_{(g)}} \left(\sigma^T_{(g)} + d\sigma^u_{(g)} - d\sigma^r_{(g)} \right) =$$

$$\frac{2}{3} \frac{d\varepsilon_{ij(g)}}{d\varepsilon_{(g)}} \sigma^T_{(g)} + \frac{2}{3} \frac{d\varepsilon_{ij(g)}}{d\varepsilon_{(g)}} \sigma^u_{(g)} - \frac{2}{3} \frac{d\varepsilon_{ij(g)}}{d\varepsilon_{(g)}} \sigma^r_{(g)}.$$

In this case, when setting the boundary value problems of plasticity and creep, the equations of equilibrium and the geometric relations are closed by the first equation, since, according to the corollary of the first theorem, it satisfies the stability condition

$$d\sigma^u_{ij(g)} / d\varepsilon_{ij(g)} > 0.$$

It can be shown that from the equation (4.46), as special cases, we obtain three-dimensional models of an ideal plastic, linear- and non-linear viscous, hardening bodies and a body with a falling deformation diagram.

5

A physico-phenomenological model of plasticity at high cyclic deformation and similar cold deformation

5.1. The experimental basis of the model

Plastic forming of parts of complex geometric shapes, for example, with cold forging, is performed in 3–5 technological transitions under conditions of complex loading and non-monotonic (often cyclic) deformation. The accumulated strain intensity reaches 1–4 units [198]. The strain intensity in each transition is 0.4–0.8. In the deformation schemes used in the processes of plastic structuring of metals, for example, with equal-channel angular pressing [94, 154], deformation is observed, depending on the method of rolling the machined workpiece between the passes, cyclic or close to it (complex loading with broken trajectories). The accumulated strain intensity reaches 10 or more.

The listed features of cold alternating deformation cause the problematic nature of the problem of its mathematical modelling. Models of the classical mathematical theory of plasticity can not adequately describe this deformation. Therefore, for example, the technology of multipass cold forging has been developed for many years on the basis of empirical relationships and production experience with all the costs inherent in this method. Consequently, the problem of developing the plasticity theory of a cyclic and close to it deformation with large amplitudes and accumulated strain rates is topical.

A Physico-Phenomenological Model of Plasticity

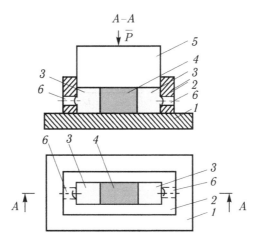

Fig. 5.1. Scheme of device and deformation of samples.

For the purpose of experimental study of the plastic behaviour of metals during cyclic deformation with large strain amplitudes, samples made from pre-annealed 10kp steel were deformed on a hydraulic press in the special device shown in Fig. 5.1. It consists of a backing plate *1*, a die *2*, and a punch *5*. Samples *4* were coated with a lubricant (flake graphite suspension in mineral oil) and deformed at room temperature in the device (flattened out under planar deformation conditions) with a punch *5*. To ensure high ductility of the steel, single-use inserts 3 made of lead were placed in the device on both sides of the sample, .

During deformation, lead was extruded into the holes *6* of the die *2*. This provided a high level of hydrostatic pressure and a sufficient plasticity of the steel in the experiment. After the first upsetting, the sample, together with the lead inserts, was pressed out from the die by a punch, if necessary it was wiped off, placed in a device with a rotation of 90°, compared to the original position, new lead inserts were placed there and the sample was subjected to a second deformation. The successive positions of the sample during the upsetting are shown in Fig. 5.2. One treatment cycle included three-fold deformation of the sample (Fig. 5.2). The accumulated intensity of the particles included two and three of the above cycles, respectively.

The samples were then cut in the direction of the largest dimension into three identical blanks from which two standard cylindrical tensile

162 Physico-Mathematical Theory of High Irreversible Strains

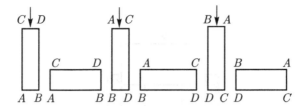

Fig. 5.2. Sequential positions of the sample in the device – one processing

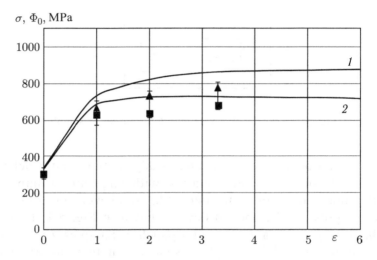

Fig. 5.3. The dependence of the intensity of stress on the intensity of the accumulated plastic deformation of steel 10 kp: the points – experiment (▲ – σ_{02} determined by tensile loading, ■ – compression); solid curves are theoretical (curve *1* – calculation according to (4.23), curve *2* – according to (5.5)).

test specimens and three cylindrical compression test specimens were manufactured. In the standard tests of the samples, yield limits were determined, which were put in correspondence with the intensity of plastic deformation accumulated by the sample during preliminary deformation of the.

Based on the results of the experimental study, the stress intensity-the strain rate $\sigma(\varepsilon)$ was constructed (Fig. 5.3, points).

According to another method, cylindrical samples with a diameter of 15 mm and a length of 80 mm made of copper M1 and aluminium AD1 were deformed under cyclic strain conditions according to the 'hourglass' pressing scheme [169, 199], as described in 3.2.1 (Fig. 3.20) .

A Physico-Phenomenological Model of Plasticity

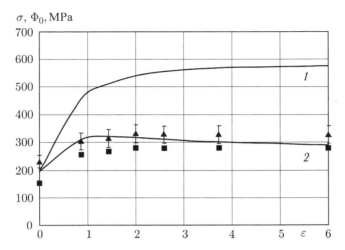

Fig. 5.4. Dependences of stress intensity on the intensity of the accumulated plastic deformation of copper M1: points – experiment (▲ – σ_{02} determination by tensile loading, ■ – compression); solid curves are theoretical (curve *1* – calculation according to (4.23), curve *2* – according to (5.5)).

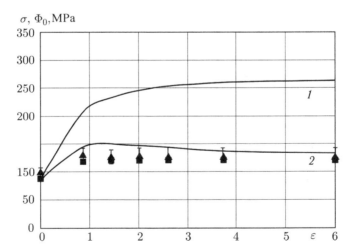

Fig. 5.5. Dependences of stress intensity on the intensity of accumulated plastic deformation of aluminum AD1: other notations as in Fig. 5.4

The stress intensity σ – the plastic deformation intensity ε curves were plotted (Figs. 5.4 and 5.5, points). Vertical segments at the points indicate a symmetrical 10-percent deviation from the mean experimental value of σ.

The dependences $\sigma(\varepsilon)$ (Figs. 5.3, 5.4 and 5.5, curves 1, respectively) were also calculated for steel 10 kp, copper M1 and aluminium AD1 for monotonic deformation by the equation of the isotropic physico-mathematical plasticity model (4.23).

Different materials and patterns of deformation of samples are chosen to determine the generality of established regularities.

The brightest of the results obtained is the fact of a significant decrease in stress intensity during cyclic deformation in comparison with monotonic deformation – the effect of cyclic (complex) loading (Figs. 5.3, 5.4 and 5.5). In comparison with the known results obtained with small deformations [29, 200, 201], the effect is more pronounced, especially for non-ferrous metals. In the case of copper and aluminium, σ decreases by 46% as compared with monotonic deformation.

In the case of cyclic deformation, as in the case of monotonic deformation, the process is stabilized (σ = const). In the first case, this occurs at values of the accumulated intensity of deformation of 1–2, in the second case, for ε = 3 (Figs. 5.3, 5.4 and 5.5). Beginning with these strains, the behaviour of materials with a high degree of accuracy corresponds to the model of an ideal-plastic body.

For metals with cyclic deformation with large strains, an insignificant anisotropy of the flow stress arises and remains constant (Figs. 5.3, 5.4 and 5.5). The difference between σ under tension and compression, referred to the average value of σ for various accumulated ε, does not exceed 16% for copper and aluminium and 11% for steel. These values correspond to the known spread of the strength characteristics of structural metals and alloys observed in their experimental determination by standard methods.

This result in the framework of the physics of strength and plasticity can be explained as follows. The preferential crystallographic orientation of the grains, which arises as a result of directional deformation in the deformation half-cycle, is accompanied by hardening, i.e., by stopping a part of the mobile dislocations by barriers and turning them into stationary dislocations. With a change in the direction of action of stresses and deformations in the next half-cycle, nothing prevents the dislocations from moving away from the barriers in the opposite direction, that is, the dislocations fixed in the original direction of deformation become mobile when the sign of deformation changes.

Thus, cyclic deformation, in comparison with monotonous, reduces the density of the stationary dislocations, that is, causes the process of softening and the cyclical 'blurring' of the deformation texture formed in the half-cycle. This is similar to the effect of elevated

5.2. The defining equations of large cyclic deformation and deformation close to it

The described experimental results served as the basis for formulating the hypothesis of a 'single curve 2' for these processes, which in a sense is analogous to the hypothesis of a single curve in the mathematical theory of plasticity [202][1].

In cyclic and close to it (complex loading with broken trajectories and trajectories of loading of large curvature) deformation of metals, characterized by high plastic strain intensities in a half cycle ($\varepsilon > 0.1$–0.2) and accumulated strains in several cycles $\int d\varepsilon \geq 1$–2, the intensity of the flow stress is a function of the intensity of the accumulated plastic strain (Udquist parameter) that does not depend on the parameters of the cycle (amplitude, symmetry, etc.) and the type of stress state.

This hypothesis makes it possible immediately to write down the defining relations in the form of the equation of the isotropic flow theory as

$$d\varepsilon_{ij} = \frac{3}{2}\frac{d\varepsilon}{\Phi_o} s_{ij}. \qquad (5.1)$$

In contrast to the well-known equation of the flow theory with isotropic hardening, in (5.1) in the denominator, instead of stress intensity σ, there is a stress function Φ_o. Figures 5.3, 5.4 and 5.5 show experimental dependences marked with dots.

In order to obtain an analytic expression for $\Phi_o(\varepsilon)$, let us turn to the physico-phenomenological model of the Bauschinger effect derived in [197], which has the form

$$\sigma = \beta m G b \left\{ \frac{\exp(\varepsilon)-1}{\lambda_c b \exp(\varepsilon)} + \frac{\rho_{so} + A\varepsilon^+}{\exp(\varepsilon)} \right\}^{1/2}, \qquad (5.2)$$

where σ, ε are the stress and strain in the case of reverse deformation after direct deformation with a power of ε^+; λ_c is the mean free path of dislocations after a change in the sign of the deformation.

---
1) The 'single curve 1' hypothesis is the hypothesis proposed by Roch and Eichinger, known in the theory of plasticity, about the independence of the deformation diagram $\sigma(\varepsilon)$ on the stress–strain state of the material within small strains.

The value of ρ_{so} for the material under investigation is determined by the method and the equation given in Section 4.3. To determine the coefficients A and λ_c a rather simple system of experiments is recommended. It involves deformation of a cylindrical specimen according to a simple stretching scheme (by drawing or direct extrusion) with an average strain of $\varepsilon^+ = 0.43$–0.6; cutting from the obtained rod three (for averaging the results) cylindrical standard samples; deformation of the samples by uniaxial upsetting with the construction of the deformation diagram $\sigma(\varepsilon)$. The values of A and λ_c are determined using the obtained diagram by formulas

$$A = \frac{(\sigma_{02}^{\exp})^2 (\beta m G b)^{-2} - \rho_{so}}{\varepsilon^+}, \qquad (5.3)$$

$$\lambda_c = \frac{b(\beta m G)^2 [\exp(\varepsilon) - 1]}{\sigma^2 \exp(\varepsilon) - (\beta m G b)^2 (\rho_{so} + A\varepsilon^+)}, \qquad (5.4)$$

where $\sigma_T^{\exp}$ is the experimentally determined yield point of the material in compression; ε and σ are the strain intensity from the interval $(0.3$–$0.5)$ and the stress intensity value corresponding to it in the experimental diagram $\sigma(\varepsilon)$.

The equations (5.3) and (5.4) are obtained from (5.2).

Since the direction of deformation changes continuously during cyclic deformation, the scalar function of the stress $\Phi_o(\varepsilon)$ describing the plastic deformation of metals under the conditions of the cyclic process under investigation is obtained from the Bauschinger effect model (5.2) by replacing ε^+ by the current value of the strain intensity ε, т е.

$$\Phi_o(\varepsilon) = \beta m G b \left\{ \frac{(\lambda_c b)^{-1} [\exp(\varepsilon) - 1] + \rho_{so} + A\varepsilon}{\exp(\varepsilon)} \right\}^{1/2}. \qquad (5.5)$$

In order to verify the consistency of model (5.5), calculations were made for steel 10kp, copper M1 and aluminium AD1 (Figs. 5.3, 5.4 and 5.5, solid curves 2). There is a very satisfactory agreement between the theoretical and experimental dependences.

Calculations were carried out at the following values of material parameters. Steel 10kp: $G = 78000$ MPa; $\lambda = 3.6 \cdot 10^{-4}$ cm; $\rho_{so} = 1.3 \times 10^{10}$ cm^{-2}; $A = 3.1 \cdot 10^{10}$ cm^{-2}; $\lambda_c = 5.4 \cdot 10^{-4}$ cm. Copper M1: $G = 46000$ MPa; $\lambda = 2.96 \cdot 10^{-4}$ cm; $\rho_{so} = 1.3 \cdot 10^{10}$ cm^{-2}; $A = 3.0 \cdot 10^{10}$ cm^{-2}; $\lambda_c = 1.18 \cdot 10^{-3}$ cm. Aluminium AD1: $G = 26000$ MPa;

$\lambda = 4.53 \cdot 10^{-4}$ cm; $\rho_{so} = 8.35 \cdot 10^9$ cm^{-2}; $A = 2.27 \cdot 10^{10}$ cm^{-2}; $\lambda_c = 1.81 \cdot 10^{-3}$ cm.

The analysis of calculated curves 2 (Figs. 5.3, 5.4 and 5.5) shows that in some cases equation (5.5) can describe diagrams of $\Phi_o(\varepsilon)$ with softening (falling diagrams), i.e. for $\varepsilon > 1$ $d\Phi_o/d\varepsilon < 0$. This can cause known mathematical difficulties in solving practical problems of studying the stress–strain state of blanks in forming operations of cold volumetric stamping. As already noted, for given ε, the plastic behaviour of materials with an accuracy that is acceptable for technological calculations[1] corresponds to the ideal-plastic body model. Therefore, it seems expedient for the formulation and solution of boundary-value plasticity problems in the first stage of deformation characterized by hardening (Figs. 5.3, 5.4 and 5.5), use the defining relations (5.1) and (5.5). In the second stage (Φ_o = const), we proceed to the formulation of the boundary value problem within the framework of the theory of ideal plasticity using the plasticity condition

$$\left(\frac{3}{2} s_{ij} s_{ij}\right)^{1/2} = \sigma_T^o, \tag{5.6}$$

where $\sigma_T^o = \Phi_o(\varepsilon) = $ const and the defining relations of the form

$$d\varepsilon_{ij} = \frac{3}{2} \frac{d\varepsilon}{\sigma_T^o} s_{ij}. \tag{5.7}$$

In this case, the duration of the first stage and the value of σ_T^o are easily determined from the dependence $\Phi_o(\varepsilon)$, constructed for the material under investigation by the equation (5.5).

It is known that the equations of the flow theory (5.1) and the theory of ideal plasticity (5.6), (5.7) ensure obtaining of satisfactory results in solving practical problems for determining the stress-strain state of blanks in the forming operations of metal working under simple loading and monotonic deformation conditions. Therefore, the fact of a satisfactory description of the experimental dependences by the scalar equation (5.5) is sufficient for their use in the mathematical modelling of the processes.

Thus, according to the foregoing hypothesis and the model presented, the behaviour of real metals in cyclic deformation with large strains in half-cycles and accumulated over several cycles can

---

1) In the theory and technology of metal forming, the accuracy of $\pm(10-15)\%$ is considered satisfactory.

be put in correspondence in the stress intensity – strain intensity coordinates the behavior of some abstract isotropic material under simple loading and monotonous deformation, the deformation diagram of which $\Phi_o(\varepsilon)$ is determined taking into account some parameters of cyclic deformation of a real metal.

6

Physico-phenomenological models of irreversible strains in metals

6.1. Model of evolution of a microstructure under irreversible deformation of metals

In contrast to the mathematical theories of plasticity and creep, the developed theory makes it possible in describing the processes of deformation in parallel with the calculation of the characteristics oif the stress–strain state to determine, using the equations (4.12) and (4.16), the scalar dislocation density $\rho^{1)}$ accumulated by the material particle in question under the conditions of warm and hot deformation in g loading steps or using equation (4.22) in monotonous cold deformation and simple loading, and to estimate the average linear grain size and subgrain by the relationship known in metals physics [192]

$$D = B/\sqrt{\rho}, \qquad (6.1)$$

where B is of the order of 10.0.

The equation for determining the scalar dislocation density in processes of large cyclic and cold deformation that is close to it follows from (4.21) and (5.5) in the form

---

1) Since $t_g = v_{gs}^{-1} \ll t_s = v_{sg}^{-1}$, where t_g is the travelling time of the mobile dislocation in the path λ, t_s is the time of 'waiting' for a fixed dislocation for the thermomechanical activation act to overcome the barrier, then $\rho_s \gg \rho_g$ and $\rho = \rho_g + \rho_s \cong \rho_s$, where ρ is the total dislocation density.

$$\rho_s = \frac{(\lambda_c b)^{-1}[\exp(\varepsilon)-1+\rho_{so}+A\varepsilon]}{\exp(\varepsilon)}. \qquad (6.2)$$

The theory also makes it possible to determine the scalar density of microcracks N_m accumulated by the particle under g loading steps using the equations (4.25), (4.39) and (4.26) and predict the probability of its macrofailure by the criteria (4.27) and (4.28). Evaluation of the characteristics of the structure of the metal in the process of deformation is necessary both in the development of technology for the production of ultrafine-grained metals, and in the development of technology for pressure processing of metals (cold, warm and hot stamping). Knowledge of the characteristics of the structure makes it possible to predict the quality and properties of the metal after deformation.

6.2. Kinetic physical-phenomenological model of dislocation creep, controlled by thermally activated slip of dislocations

Modern computational methods of the mathematical creep theory, based on the technical floe theories, hardening, ageing and heredity theory, were developed on the basis of the fundamental results obtained in the 1960s and 1970s [41]. The method of constructing the phenomenological creep theory is analogous to the method of constructing the mathematical theory of plasticity. Therefore, it has the same drawbacks.

Microstructural uniaxial creep models have been developed for many years independently of the phenomenological theory. A detailed survey of them is given in monographs [44, 45] and was analyzed in section 3.2.4. Different authors have proposed for several materials and loading conditions several models that differ in the initial assumptions and contain a large number of fitting parameters and do not describe non-stationary creep. Therefore, they practically can not be used in solving applied problems.

The problem of creep is, in the main, the problem of viscoplasticity. Therefore, it can be assumed that, on the basis of the above model of viscoplasticity we can obtain the basic equations of dislocation creep, controlled by thermodynamically activated gliding of dislocations.

Just note that the model below does not pretend to be universal. In this paper, we construct models of irreversible strains that can be used in engineering calculations. The results of their verification

Physico-Phenomenological Models of Irreversible Strains 171

during deformation of specific metals are given. The actual ranges of applicability of models can be established only as a result of numerous studies carried out on different metals and alloys.

To ensure that the load history can be taken into account, the creep model, like the plasticity model, will be constructed in finite increments.

To derive equations describing the creep curve $\varepsilon^c(t)$, we assume that at an arbitrary calculated loading step g, characterized by a small finite time increment, $dt_{(g)}$, $\dot{\varepsilon}^c_{(g)}$ and $T_{(g)}$, because of the smallness of $dt_{(g)}$, do not change significantly, i.e/ $\dot{\varepsilon}^c_{(g)}$, $T_{(g)}$ = const, but they can take different values at different calculated steps g. Instantaneous at an arbitrary design step g, the rate of creep strain rate for given σ and T is determined from the original equation (4.3) as

$$\dot{\varepsilon}^c_{(g)} = \dot{\varepsilon}_* b \sqrt{\rho_{s(g-1)}} \exp\left(-\frac{\beta(T)G(T)b^3 - \sigma^c b^2/m\sqrt{\rho_{s(g-1)}}}{kT}\right), \qquad (6.3)$$

where the superscript c here and below denotes creep macro-characteristics, and $\dot{\varepsilon}_* b\sqrt{\rho_{s(g-1)}} = \dot{\varepsilon}_0$.

To introduce the main time variable into the model, we substitute $d\varepsilon^c_{(g)} = \dot{\varepsilon}^c_{(g)} \cdot dt_{(g)}$ into the equation (4.12) and the change in the density of the stationary dislocations in creep under the applied stress intensity $\sigma^c_{(g)}$ on step g in time $dt_{(g)}$ we obtain in the form

$$d\rho_{s(g)} = \left[\frac{\dot{\varepsilon}^c_{(g)}}{b\lambda} - (\rho_{s(g-1)})^{3/2} v_D b \times \exp\left(-\frac{\beta(T)G(T)b^3 - \sigma^c_{(g)}b^2/m\sqrt{\rho_{s(g-1)}}}{kT}\right)\right]dt_{(g)}. \qquad (6.4)$$

The intensity of the creep strain increment for $dt_{(g)}$ with allowance for (6.3) is determined from equation

$$d\varepsilon^c_{(g)} = \dot{\varepsilon}^c_{(g)} \cdot dt_{(g)}. \qquad (6.5)$$

The intensity of creep strain, accumulated over g steps, can be found from the formula

$$\varepsilon^c_{(g)} = \varepsilon^c_{(g-1)} + d\varepsilon^c_{(g)}, \qquad (6.6)$$

and the time for the sample to remain under load for g calculated steps – according to the formula

$$t_{(g)} = t_{(g-1)} + dt_{(g)}. \tag{6.7}$$

The $\varepsilon^c_{(g)}$ found from (6.6) is associated with the time $t_{(g)}$ according to (6.7).

Further, taking into account the $d\rho_{s(g)}$ determined by (6.4), the dislocation density is found at an arbitrary calculated step g:

$$\rho_{s(g)} = \rho_{s(g-1)} + d\rho_{s(g)}. \tag{6.8}$$

The value of $\rho_{s(g)}$ determined from (6.8) is substituted in (6.3), and the strain rate intensity is found at the next calculated step $(g + 1)$, and the calculated cycle involving formulas (6.4)–(6.8) is repeated. For $g = k$ calculated steps (cycles), a theoretical creep curve $\varepsilon^c(t)$ is constructed. If at some instant $t(g)$ creep modes change, that is, T, σ^c = var, then the corresponding calculated step g in the above formulas must be substituted for their new values. The intensity of accumulation ρ_s changes. This real structural parameter in theory also ensures that the loading history is taken into account in step-by-step calculation.

When calculating the creep curves, as well as the deformation diagrams, small increments of time $dt_{(g)}$ and the intensity of the deformation increment $d\varepsilon_{(g)}$ must be specified from the condition

$$dt_{(g)} = d\varepsilon_{(g)} / \dot{\varepsilon}_{(g)} \gg t_h = v_{so}^{-1} =$$
$$\left(v_D b \sqrt{\rho_s}\right)^{-1} \exp\left(\frac{\beta G b^3 - \sigma b^2 / m \sqrt{\rho_s}}{kT}\right), \tag{6.9}$$

where t_h is the characteristic time of the process. If $dt_{(g)} = \dfrac{d\varepsilon_{(g)}}{\dot{\varepsilon}_{(g)}} < t_h$ dislocations do not have time to overcome barriers and creep will not be observed during this time. In the same way, if $\dot{\varepsilon}_{(g)} = \dfrac{d\varepsilon_{(g)}}{dt_{(g)}} > t_h^{-1}$ the process of disappearance of the stationary dislocations will not have time to occur and the deformation, despite the high temperature, will be cold (it will be determined by the hardening process).

From the scientific and practical point of view of great interest is the steady-state (stationary) creep flowing at T, with a minimum strain rate $\dot{\varepsilon}^c_{min} = d\varepsilon^c / dt = const$. In creep physics, it is considered that the necessary condition for steady creep is the stationarity condition of the structure [44, 45]. In the presented creep model, this condition has the form $d_{ps}/dt = 0$. By substituting it in (6.4), we find $\dot{\varepsilon}_{min}^c$ with which the steady-state creep takes place, in the form[1])

$$\dot{\varepsilon}_{min}^c = \lambda \rho_s^{3/2} v_D b^2 \exp\left(-\frac{\beta(T)G(T)b^3 - \sigma^c b^2/m\sqrt{\rho_s}}{kT}\right). \tag{6.10}$$

Equating (6.3) and (6.10), we find the equation for determining the stationary dislocation density ρ_s^c at the steady-state creep stage for a particular material:

$$\lambda \rho_s^c b = \dot{\varepsilon}_*/v_D = C = \text{const.} \tag{6.11}$$

If $\dot{\varepsilon}_* = v_D$, as is customary in this paper, then C in (6.11) is of the order of 1.0. If we take into account that v_D for different metals has values 10^{12}–10^{13} s^{-1} [67], then the values of C for different metals can vary within the limits of 0.1–1.0.

Expression (6.11) is a mathematical formulation of a new (not known earlier) structural law of dislocation creep: *with the steady-state dislocation creep of metals with thermomechanical activation of dislocation gliding, the product of three scalar characteristics of the dislocation structure: the mean free path of dislocations λ, the scalar dislocations ρ_s and the density of the Burgers dislocation vector b is for a given metal a constant value independent of σ^c, T and cumulative ε^c.*

The fact that steady-state creep occurs under conditions of a stationary structure was assumed earlier [44, 45], and in this paper the condition $d_{ps}/dt = 0$ was adopted. The product $\lambda \rho_s^c b$ itself is non-trivial in (6.11) and it independent of the thermal and speed conditions, which justifies the characteristic (6.11) as a law.

Since in the estimates $b = 3 \cdot 10^{-8}$ cm $=$ const, the mean free path λ and the dislocation density ρ_s at the steady state are uniquely related.

The law (6.11) allows:

1. Estimate the steady-state density of dislocations with steady-state creep as

---

[1]) In the equations concerning steady-state creep, the index g is omitted.

$$\rho_s^c = C/\lambda b. \tag{6.12}$$

2. Find the equation for determining the minimum strain rate for steady-state creep. This equation is obtained by substituting (6.12) into (6.10) and has the form

$$\dot{\varepsilon}_{\min}^c = \lambda \left(\frac{C}{b\lambda}\right)^{\frac{3}{2}} \cdot v_D b^2 \exp$$
$$\left(-\frac{\beta(T)G(T)b^3 - \sigma^c b^2 \cdot \sqrt{b\lambda}/m\sqrt{C}}{kT}\right). \tag{6.13}$$

3. Experimental studies of the effect of creep strain on the structure of metals have established that a general regularity is the formation of a subgrain structure at the beginning of the steady-state stage at T, σ^c = const. To estimate the average linear size of subgrains, an empirical relationship has been proposed [45]

$$D = 10.5 Gb/\sigma^c. \tag{6.14}$$

The equation for estimating the average linear size of subgrains in the framework of the proposed theory is derived by substituting (6.12) in (6.1):

$$D = B/\sqrt{\rho} \cong B/\sqrt{C/b\lambda}. \tag{6.15}$$

However, the question of the value of C in (6.11) in the case of alloys requires an independent systematic study.

It can be shown that the above scalar physico-phenomenological creep model (equations (6.3)–(6.8)) allows the construction of technical physical and mathematical creep theories of metals using the associated flow law: the theory of aging $\sigma^c = \Phi_1(\varepsilon^c, t)$, the flow theory $\sigma^c = \Phi_2(\dot{\varepsilon}^c, t)$, the hardening theory $\sigma^c = \Phi_3(\dot{\varepsilon}^c, \varepsilon^c)$ [7]. Let's show it on the example of the theory of agring, which is widely used in engineering calculations.

In the mathematical theory of ageing, the creep potential is taken in the form

$$f_1 = \frac{3}{2} s_{ij}^c \cdot s_{ij}^c - \left[\Phi_1(\varepsilon^c, t)\right] = 0. \tag{6.16}$$

It follows from (6.16) that the relationship between σ_{ij}^c and ε_{ij}^c at some time t is determined by the equation [7]

$$\varepsilon^c_{ij} = \frac{3}{2}\frac{\varepsilon^c}{\sigma^c}(\sigma^c_{ij} - \delta_{ij}\sigma^c_o) = \frac{3}{2}\frac{\varepsilon^c}{\sigma^c}s^c_{ij}, \qquad (6.17)$$

which is supplemented by the scalar relation $\sigma^c = \Phi_1(\varepsilon^c, t)$.

Dependences $\sigma^c = \Phi_1(\varepsilon^c)$, constructed for different values of time t, are called *isochronous creep curves*. They are constructed by processing a series of experimental creep curves obtained for a specific T and different σ [7]. In this case, the stresses and strains in creep in calculating the parts for a particular time value are found by solving the initial boundary value problem, the mathematical formulation of which includes equations (6.16) and an isochronous creep curve for the time t of interest, i.e., the solution is equivalent to solving the deformation theory of plasticity using the isochronous creep curve for the time of interest instead of the material deformation diagram.

The uniaxial physico-phenomenological creep model makes it possible to calculate the necessary number of creep curves for various T and σ^c, which are necessary for constructing isochronous curves.

The physico-phenomenological technical creep theories constructed in this way (using the developed scalar creep model) will differ from the phenomenological ones by the smaller volume of the experiment in order to obtain isochronous creep curves in solving practical problems related to the calculation of creep parts. However, it is quite obvious that these theories will inherit the main drawback of mathematical theories of creep: the impossibility of correctly taking into account the loading history and, therefore, an adequate description of the unsteady creep process.

The problem of creep is the problem of viscoplasticity. Because of the above the physico-mathematical theory of viscoplasticity, irreversible deformation is not divided into plastic deformation and creep deformation. The scalar physico-phenomenological creep model is derived from the scalar model of viscoplasticity. Therefore, the solution of creep problems in the framework of the physical and mathematical theory is equivalent to solving problems of viscoplasticity. We shall call them the problems of the *physico-mathematical theory of irreversible strains*.

Consequently, the mathematical formulation of the initial-boundary problems of irreversible strains should include equations the (4.12)–(4.20) and (4.46). If it is necessary to predict the probability of fracture or a service life (long-term strength), the equations (4.25)–(4.28) are added to the indicated equations. In parallel with the definition of the characteristics of the stress–strain state, it is necessary to record the time of finding the material particles under the load by the equation (6.7).

For a more accurate determination of the characteristics of the stress–strain state and deformation damage in irreversible deformation processes it is expedient, in addition to the experimental production of a 'cold' material deformation diagram, which determines the characteristics of the initial dislocation structure, ρ_{so} and λ, to have two or three experimental deformation diagrams or two or three experimental creep curves of the material for more accurate 'tuning' of models. The adjustment of the models reduces to a more accurate determination of the dependences of $\beta(T, \sigma^c)$ for the material under study.

6.3. Kinetic physico-phenomenological model of long-term strength of metals

6.3.1. General information about long-term strength

The strength characteristic of a material that is in a stressed state for a long time at a high temperature is the limit of long-term strength. *The limit of long-term strength is the conditional stress, that is, the ratio of the tensile force to the original area of its cross-section, at which the sample breaks down after a certain time interval* [7].

According to this definition, the long-term strength characterizes the working capacity of the material under creep conditions at a certain temperature and stress.

Macroscopic failure in creep, that is, in the case of irreversible deformation, which depends on time and proceeds at a stress below the yield point, as in the case of active deformation, can be ductile (occurs with great deformation with neck formation) and brittle. Ductile fracture during creep has intragranular (transcrystalline) character, and brittle – intergranular (intercrystalline) character.

Ductile fracture, as a rule, is observed at relatively low temperatures and high voltages (high $\dot{\varepsilon}$). Embrittlement of the material occurs with a long stay at high temperatures, that is, at low voltages.

From the definition of the long-term strength limit it follows that its value for a particular material depends on temperature and time until the moment of macrofracture. The time to fracture is chosen equal to the service life of the part. The temperature is determined by the operating conditions of the part. The load that is selected will lead to the appearance of a stress under the action of which the part will fail after a selected period of time at a given temperature

Typically, the time base for performing design calculations ranges from 100 to 100 000 hours. It also follows from the definition that with increasing temperature and a given operating time of the part, the limit of long-term strength decreases.

In the experimental determination of the long-term strength of the material, several identical samples are tested at different stresses at the same temperature. The time to fracture of each sample is recorded. According to the established results, a plot is constructed of the dependence of the long-term strength limit on the time to fracture of the sample, and the limit of the long-term strength for a given period of time is determined from it.

Carrying out test data at different temperatures, we find the limits of long-term strength at different temperatures and times.

The main dependence in the theory of long-term strength is the dependence of the long-term strength limit on time to failure in Fig. 6.1. In logarithmic coordinates, the dependence has the form of a broken line. The inflection point of the graph, as a rule, corresponds to the transition from transcrystalline to intercrystalline fracture.

The dependence of time to fracture from on the limit of long-term strength for a particular material at a certain temperature is a power dependence:

$$t = A_1 \sigma_{\text{long}}^{-m}, \qquad (6.18)$$

where the material constants A_1 and m depend on T and the nature of fracture and, apparently, on the structure.

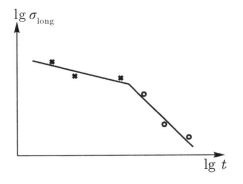

Fig. 6.1. Dependence of the limit of long-term strength on time to fracture at $T = $ const [7].

In early studies of long-term strength at non-stationary temperatures and stresses, the law of linear summation of damage was established

$$\sum_{i=0}^{k} \frac{t_i^*}{t_{i\,\text{frac}}} = 1.0, \qquad (6.19)$$

where $i = 1, 2, ..., k$ is the number of the test phase of the sample, in which T and σ have constant values for the time t_i^* and change as you move to the next stage; t_i is the time to failure at T and σ constant for each stage.

With a continuous stress change

$$A = \int_0^{t_{\text{frac}}} \frac{dt}{t(\sigma)_{\text{frac}}} = 1.0. \qquad (6.20)$$

Later it was found that often $A \neq 1.0$ [119]. This result is more natural than (6.20), since fracture is an unbalanced process and must depend on the loading history. The *criterial and kinetic approaches* are currently used [119] in describing the long-term strength under the conditions of a complex stress state.

As was shown in Chapter 4 in the construction of physico-phenomenological theories of viscoplasticity and ductile fracture, one can consistently take into account the loading history (taking into account transient processes with changing T and σ) only if there are kinetic equations in the theories for the actual structural parameters of metals and constructing the theory in finite increments.

In the first case, the accuracy of predicting the long-term strength of parts depends, as in calculations for static strength, on the successful choice of the expression for the equivalent stress σ_e. Four basic combinations of principal stresses were considered as the equivalent stresses: the maximum principal stress σ_{e1}, the stress intensity σ_{e2}, their half-sum σ_{e3}, the difference between the maximum and minimum principal stress σ_{e4} [119], etc.

This approach is used in analyzing the long-term strength under steady-state loading conditions, when the components of the stress tensor do not change with time.

In [203], based on the analysis of the proposed long-term strength criteria, it was shown that in describing the experimental data the best results are shown by the power-law model

$$t = C\sigma_e^{-n} \qquad (6.21)$$

and the fractional-power type

$$t = D\left[\left(\sigma_B - \sigma_e\right)/\sigma_e\right]^m, \qquad (6.22)$$

where σ_e is the limit of short-term strength at the test temperature.

In the kinetic approach, different versions of the theory are distinguished by the choice of the measure of damage, which can be a scalar, a vector, a tensor, or a collection of, for example, scalar and vector parameters [204].

Since the beginning of the 1970s, attempts have been made in a number of works [205, 206, 207] to relate the macroparameter of damage to the characteristics of the metal structure. The main emphasis was placed on the description and analysis of the effect of micropores, which, as the experiment testified, arose at the grain boundary. However, this direction has not been properly developed.

The present state of the phenomenological theories of creep and long-term strength is described in more detail in the reviews mentioned above [52, 53, 119, 205]. Here, on the basis of a brief discussion of the question, we make the following conclusion.

1. The problem of long-term strength, which is closely related to the problem of creep, has been the subject of research by mechanics for many years. A lot of theoretical and experimental material has been accumulated. Several variants of the description of the process in the framework of both the criterial approach and the kinetic approach have been proposed.

2. The main task for today is a wide experimental study and a theoretical description of the long-term strength under conditions of a complex and unsteady state of stress and thermal action.

3. A prospective approach to generalizing the accumulated results and solving the formulated problem is the physico-phenomenological approach developed in this book. In this case, the essence of this approach is the combination of the above physico-phenomenological models of creep and ductile fracture of metals. Since within the framework of this approach the actual structural parameters (scalar dislocation and microcrack density) and physical kinetic equations for them are introduced into the theory, the model constructed in final increments should correctly describe the long-term strength of metals under the conditions of complex and non-stationary thermomechanical loading.

6.3.2. Model of long-term strength. The general case of loading

It is obvious that a physico-phenomenological model of long-term strength can be obtained by synthesis of creep models (Section 6.2) and ductile fracture (Section 4.4) of metals [208].

Let's consider the general case of loading of a part from a concrete material in the conditions of non-stationary temperature and the stress–strain state. We will formulate the model in finite increments, assuming that for small increments of time at the calculated step g at different parts of the workpiece with the coordinates x_k, y_k, z_k, $k = 1, 2, ..., n$, the temperature and the stress tensor components do not have time to change significantly, that is, $T_{(g)}$ and $\sigma_{ij(g)}$ = const.

At the next calculated step $(g + 1)$ at each point of the volume of the part, $T_{(g+1)} \neq T_{(g)}$ and $\sigma_{ij(g+1)} \neq \sigma_{ij(g)}$, but their values are known from the parallel solution of the thermal and initial-boundary tasks. Consequently, the long-term strength equations described below refer to each point of the calculated part. It is clear that the calculation by the final increments dt is divided into n-th number of steps.

In the long run, the long-term strength model will have the form of a scalar functional

$$t^* = F\left[T(t), \sigma^c(t), \varepsilon^c(t), \rho(t), N_m(t), dt\right], \qquad (6.23)$$

where t^* is the time through which at some point of the part there will be a macrofracture, i.e., the condition (4.28) will be satisfied; $T(t)$, $\sigma^c(t)$, $\varepsilon^c(t)$, $\rho(t)$ and $N_m(t)$ are time functions describing the change in temperature, intensity of stress, intensity of accumulated irreversible strain, dislocation density and microcrack density, respectively, at the point where there was a macrofracture. In principle, these functions can be constructed tabularly or in the form of a graph after the end of the calculation, using its results[1].

Instantaneous at an arbitrary design step g, the creep strain rate at σ^c, T and ρ_s, known at this step, is determined from equation (6.3):

$$\dot{\varepsilon}^c_{(g)} = \dot{\varepsilon}_* b \sqrt{\rho_{s(g-1)}} \exp\left(-\frac{\beta(t)G(t)b^3 - \sigma^c b^2/m\sqrt{\rho_{s(g-1)}}}{kT}\right). \qquad (6.24)$$

The intensity of the creep strain increment over time $dt_{(g)}$ is found from equation (6.5):

---

[1] The novelty of the theory developed in the work can be drrn by comparing equation (6.23) with equations (6.21) and (6.22).

Physico-Phenomenological Models of Irreversible Strains 181

$$d\varepsilon^c_{(g)} = \dot{\varepsilon}^c_{(g)} dt_{(g)}. \tag{6.25}$$

The change in the dislocation density at the step g during the time $dt_{(g)}$ is determined from equation (6.4):

$$d\rho_{s(g)} = \left[\frac{\dot{\varepsilon}^c_{(g)}}{b\lambda} - \left(\rho_{s(g-1)}\right)^{3/2} v_D b \times \exp\left(-\frac{\beta(T)G(T)b^3 - \sigma^c_{(g)}b^2/m\sqrt{\rho_{s(g-1)}}}{kT_{(g)}}\right) \right] dt_{(g)}. \tag{6.26}$$

The intensity of creep strain, accumulated over g steps, is found as

$$\varepsilon^c_{(g)} = \varepsilon^c_{(g-1)} + d\varepsilon^c_{(g)}. \tag{6.27}$$

The dislocation density accumulated at a point beyond the calculated steps is determined by the formula (6.8):

$$\rho_{s(g)} = \rho_{s(g-1)} + d\rho_{s(g)}. \tag{6.28}$$

The time of the considered microvolume of the part under the The weight of the calculated steps is found from formula

$$t_{(g)} = t_{(g-1)} + dt_{(g)}. \tag{6.29}$$

By setting the time $t_{(g)}$ (6.29) in correspondence with $\varepsilon^c_{(g)}$ (6.27), one can construct a non-stationary creep curve for the microvolume of the part under consideration until the moment of failure.

The change in the density of microcracks in the increment of deformation (6.25) is calculated from equation

$$dN_{m(g)} = \left[\xi_0 \rho_s \frac{v_D b \sqrt{\rho_{s(g)}}}{\dot{\varepsilon}^c_{(g)}} \exp\left(-\frac{\beta_s(T)G(T)b^3 - \sigma^c_{(g)}b_2/m\sqrt{\rho_{s(g)}}}{kT_{(g)}}\right) - N_{v(g-1)} \frac{v_D b}{\pi \bar{\xi} \dot{\varepsilon}^c_{(g)}} \exp\left(-\frac{\beta(T)G(T)b^3 + \sigma^c_{(g)} K_{(g)} \pi \bar{\xi} b^2/mM}{kT_{(g)}}\right) \right] d\varepsilon^n_{(g)}. \tag{6.30}$$

182 *Physico-Mathematical Theory of High Irreversible Strains*

The density of microcracks accumulated by the microvolume for g calculation steps and, consequently, for the time $t_{(g)}$ (6.29) is determined by the formula (4.26)

$$N_{m(g)} = N_{m(g-1)} + dN_{m(g)}. \quad (6.31)$$

The conclusion about the probability of macrofracture is made using the criteria (4.27) and (4.28):

$$\psi_{(g)} = N_{m(g-1)} / N^*_{(g)} < 1,0, \quad (6.32)$$

$$\psi_{(g)} = N_{m(g-1)} / N^*_{(g)} = 1,0. \quad (6.33)$$

If the macrofracture criterion (6.33) is satisfied for a given workpiece point, then the time value $t_{(g)} = t^*$ (6.29) at which this occurs is the work life of the workpiece under the given thermomechanical conditions.

From the foregoing model of long-term strength it follows that for its use in design calculations, it is necessary to develop a computer program for mathematical modelling and solving problems in the physical and mathematical theory of the strength and plasticity of metals.

6.3.3. Modeling of the process of testing samples for long-term strength under conditions of stationary thermomechanical loading

Since in this case the stress and temperature do not change during the testing of each sample, it is expedient to formulate the model in the final equations. To do this, we adopt the following simplifying assumptions.

1. The stress state of a sample loaded with a tensile stress $\sigma = P/F_0$, where P = const is the tension force; $\sigma < \sigma_{02}$; F_0 is the initial cross-sectional area of the sample, is linear. This assumption is justified, since, by definition, the long-term strength limit σ_{long} is a conditional stress.

2. We assume that most of the time (up to the moment of destruction) the sample is deformed under the conditions of stationary creep. This assumption has an experimental justification and avoids the mathematical difficulties associated with the integration of a complex non-linear differential equation (6.30).

Physico-Phenomenological Models of Irreversible Strains

Thus, in equation (6.30): $\rho_s = C/\lambda b$ (on the basis of the structural law of stationary creep (6.11)); $K = \sigma_0/\tau$; $\sigma_0 = \sigma/3$; $\tau = \sigma/\sqrt{3}$. The value of λ is determined from the 'cold' material deformation diagram, as described in section 4.3.

Going to equation (6.30) with the help of the dependence $d\varepsilon^c = \dot{\varepsilon}^c dt$ to the main variable – time t, after its integration with the initial conditions $t = 0$, $N_m = N_{m0}$ is the initial scalar density of microcracks in the material, we obtain

$$t^* = \frac{1}{q}\ln\frac{|n - N_{m0}q|}{|n - N_m^* q|},\qquad(6.34)$$

where is denoted:

$$q = \frac{v_D b}{\pi \bar{\xi}}\exp\left(-\frac{\beta(T)G(t)b^3 + \sigma_{\text{long}}K\pi\bar{\xi}b^2/mM}{kT}\right);\qquad(6.35)$$

$$n = \xi_0 \rho_s v_D b\sqrt{\rho_s}\exp\left(-\frac{\beta(T)G(t)b^3 + \sigma_{\text{long}}K\pi\bar{\xi}b^2/m\sqrt{\rho_s}}{kT}\right);\qquad(6.36)$$

$N_m^* = 10^6$ cm^{-2} is the critical density of the microcrack during deformation by uniaxial ($K = 0.58$)[1] stretching; t^* is the time after which the sample, loaded with tensile stress σ_{long} at a given temperature T, is destroyed – long-term strength; σ_{long} is the long-term strength limit.

6.4. Stress relaxation model

It is not difficult to understand that the viscoplasticity theory proposed in Ch. 4 can be generalized to an elastoviscoplastic medium by adding to the equations (4.46) an elastic component of deformation in the form of Hooke's law:

$$d\varepsilon_{ij(g)} = \frac{3}{2}\frac{d\varepsilon_{(g)}}{\sigma_{(T)}^T + d\sigma_{(g)}^u + d\sigma_{(g)}^r}\left(s_{ij(g)}^T + ds_{ij(g)}^u + ds_{ij(g)}^r\right) + 1/2G\left(d\sigma_{ij(g)} - \delta_{ij}\frac{3\nu}{1+\nu}d\sigma_{o(g)}\right). \qquad(6.37)$$

Since the irreversible strain in the viscoplasticity model (4.46) is

---
1) The values of N_m^* for various K are given in ection 7.2.

not subdivided into plastic strain and creep strain, $d\varepsilon_{(g)}$ in (6.37) is elastoviscoplastic strain.

Let us create a physico-phenomenological model of viscoelasticity–stress relaxation in a certain load loaded in the elastic range and a fixed body.

If the material particle initially is loaded with an intensity stress at the calculated step $g = 1$, characterized by a small but finite time increment $dt_{(g)}$, then it will receive an elastic deformation of the intensity equal to $\sigma_{(g)} < \sigma_{02}$,

$$\varepsilon^e_{(g)} = \sigma_{(g)}/E, \qquad (6.38)$$

where E is the Young's modulus, the superscript e here and below will mean elastic strain. For $g = 1$, $\sigma_{(1)} = \sigma_0$ is the initial stress.

During the time $dt_{(g)}$, part of the elastic deformation goes into an irreversible creep deformation of the intensity (6.5) with the velocity (6.3). The remaining elastic deformation will be equal to

$$\varepsilon^e_{(g+1)} = \varepsilon^e_{(g)} - d\varepsilon^c_{(g)}. \qquad (6.39)$$

This strain will correspond to the stress

$$\sigma_{(g+1)} = E\varepsilon^e_{(g+1)}. \qquad (6.40)$$

The dislocation density in the time $dt_{(g)}$ changes by an amount

$$d\rho_{s(g)} = \left[\frac{\dot{\varepsilon}^c_{(g)}}{b\lambda} - \left(\rho_{s(g-1)}\right)^{3/2} v_D b \times \exp\left(-\frac{\beta(T)G(T)b^3 - \sigma_{(g)}b^2/m\sqrt{\rho_{s(g-1)}}}{kT_{(g)}}\right)\right]dt_{(g)}. \qquad (6.41)$$

The dislocation density at the calculated step $(g + 1)$ is equal to

$$\rho_{s(g+1)} = \rho_{s(g)} + d\rho_{s(g)}. \qquad (6.42)$$

In this case, the relaxation time per g steps

$$t_{(g)} = t_{(g-1)} + dt_{(g)}. \qquad (6.43)$$

This completes the first calculation cycle, and the next on $(g + 1)$

is again based on the definition of $\dot{\varepsilon}^c_{(g+1)}$ by (6.3).

The calculation ends when, in (6.38), $\varepsilon^e_{(g)} \to 0$, i.e., the initial elastic strain almost completely changes into an irreversible creep strain.

The stress relaxation curve $\sigma(t)$ is obtained by matching the values of the time (6.43) with the stress intensity values in accordance with (6.40).

These calculations are carried out for each microvolume of the loaded body. Before beginning the calculation of stress relaxation, an elastic problem must be solved for the body in question, and the stresses and elastic strains at the points of the body are determined. This information will be the starting point for carrying out the described stress relaxation calculation using equations (6.37).

7

Experimental verification of adequacy of models

7.1. Scalar viscoplasticity model

7.1.1. Methodology for checking the adequacy of the model.

From the physico–phenomenological model of viscoplasticity, considered in section 4.2 follows the statement: *the physico-phenomenological functional of the form* $\sigma = F\left[t, \varepsilon(t), \dot{\varepsilon}(t), T(t), \rho_s(t)\right]$ *for the known deformation diagram σ(ε) of the material obtained under conditions of cold deformation allows us to determine the stress of the material flow in the temperature–velocity range in which dislocation gliding is the dominant deformation mechanism, and the main mechanism of hardening is the blocking of moving dislocations by barriers of the dislocation type (forest dislocations, grain and subgrain boundaries).*

An experimental verification of this statement and, consequently, of the uniaxial physico-phenomenological model of plasticity and viscoplasticity was carried out by comparing the experimental and theoretical diagrams of the deformation of metallic materials.

In order to assess the generality and range of applicability of the model, the study was conducted with the involvement of a sufficiently wide range of metals and alloys, which differ in the type of crystal lattice, the basis, and the chemical composition. The deformation diagrams constructed and calculated in sufficiently wide temperature and strain rate ranges were subjected to comparison. The experimental deformation diagrams of high-quality steels 10, 30G1P, 20G2P and 12Cr18Ni9 stainless steel were used, which were specially obtained within the framework of this work using a 1231U-10 testing machine, as well as diagrams of various steels and alloys published in the scientific and technical literature.

The initial information is a diagram of the cold deformation of the investigated material. Using this diagram, $\sigma(\varepsilon)$ from the equations given in section 4.3, the characteristics of the initial dislocation structure of the material ρ_{so} and λ were determined. We note that the initial dislocation density ρ_{so}, determined by the recommended technique for different materials, corresponds in the order of magnitude to the experimental data available in the literature (for example, in steels after hot rolling $\rho_{so} = 10^9 - 10^{10}$ cm^{-2}) [67].

The following values of the model constants were used: $b = 3 \times 10^{-8}$ cm — the modulus of the Burgers dislocation vector averaged over the slip systems; $m = 3.1$ — Taylor factor for polycrystals with chaotic misorientation of grains; $k = 1.38 \cdot 10^{-23}$ J/K is the Boltzmann constant; $v_D = 10^{12}$ s^{-1} is the Debye frequency. Also for calculation it is necessary to know the values of the shear modulus G of the material at different temperatures. If the value of the Young's modulus E for the material is known, then G was determined as $E/2(1 + v)$, where $v = 0.325$ is the average Poisson's ratio for the alloys. The values of $E(T)$ and $G(T)$ were taken from [209].

In the first step, substituting in (4.23) the values found for G, ρ_{so} and λ, the value of λ was corrected, achieving the best coincidence of the 'cold' experimental and calculated (4.23) dependences $\sigma(\varepsilon)$. After that, we began to calculate the material deformation diagrams for different values of T and $\dot\varepsilon$ and for various laws of their variation during deformation.

The values of the model parameters for high-quality carbon steels and 12Cr18Ni9 stainless steel are given in Table. 7.1.

Theoretical deformation diagrams were obtained by mathematical modelling of the upsetting of cylindrical quasi-samples under uniaxial

Table 7.1. The values of the model parameters used to calculate the deformation diagrams

Steel grade	$\rho_{so} \cdot 10^{10}$, cm^{-2}	$\lambda \cdot 10^{-4}$, cm	G (20°C), MPa	G (450°C), MPa	G (650°C), MPa	G (850°C), MPa	σ_T, MPa
10	1.3	4.7	78 000	75 000	51 000	53 400	332.2
30G1R	2.64	2.2	77 270	69 500	46 500	52 000	466.6
20G2R	2.08	3.3	77269.8	74 800	49 000	54 000	414.3
12Cr18Ni9	0.6	1.2	77358.5	58 000	57 800	55 000	230

stress conditions using he equations of the model (4.11)–(4.17). For this purpose, a special program was compiled in the Delphi 7 visual programming system.

The calculation of the deformation diagram was carried out in the following sequence.

At the first calculated step, $g = 1$, the calculated increment of the strain intensity $d\varepsilon(g) = 0.05$ was set. The strain rate was determined as $\dot{\varepsilon}_{(g)} = v_{\text{def}(g)}/H$, where H is the height of the virtual sample (in calculations $H = 7$ mm); $v_{\text{def}(g)}$ is the strain rate (speed of movement of the deforming organ of the testing machine). The relation $\varepsilon_{(g)} = \varepsilon_{(g-1)} + d\varepsilon_{(g)}$ was used to determine the strain rate accumulated for g steps. For $g = 1$, $\varepsilon_{(1-1)} = \varepsilon_0 = 0$ and $\varepsilon_{(1)} = 0 + d\varepsilon_{(1)} = d\varepsilon_{(1)}$. The initial flow stress $\sigma_{(1)}^T = \sigma_T$, corresponding to $\varepsilon_0 = 0$ for a rigid–plastic medium, was calculated from formula (4.13).

Next, the increment of the density of stationary dislocations $d\rho_{s(1)}$ was determined by the formula (4.12).

Then, the increment of stress intensity was calculated from formula (4.11).

The stress intensity $\sigma_{(1)}$ was calculated from the formula (4.15) – $\sigma_{(1)} = \sigma_{(1)}^T + d\sigma_{(1)}$ and put it in correspondence with the strain $\varepsilon_{(1)} = d\varepsilon_{(1)}$. Next, according to formula (4.16), the density of stationary dislocations accumulated at the step was found. For $g = 1$, $\rho_{s(1)} = \rho_{s(1-1)} + d\rho_{s(1)} = \rho_{so} + d\rho_{s(1)}$. This concludes the cycle.

Then, the second step of the loading was calculated ($g = 2$). The increment of the density of the stationary dislocations at the second step $d\rho_{s(2)}$ and so on was determined. The calculated cycles are repeated until the required value of $\varepsilon_{(g)}$ is reached.

7.1.2. Results of model verification

The results of verification of the adequacy of the model on the example of a number of high-quality steels and stainless austenitic steel are shown in Figs. 7.1–7.8. The experimental deformation diagrams $\sigma(\varepsilon)$ in the graphs are shown here and below by the points, and the theoretical ones by the solid lines. The segments at the points hereinafter denote a symmetrical ten percent deviation from the mean experimentally determined value of σ.

Analysis of these results allows us to draw the following conclusions.

1. The greatest discrepancy between the calculated values of σ and the experimental values (up to 20% of the mean experimentally

Experimental Verification of Adequacy of Models

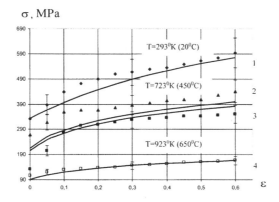

Fig. 7.1. Steel 10 deformation diagrams 10 at different temperatures T and strain rates $\dot{\varepsilon}: 1 - \dot{\varepsilon} = 10^{-2}\,\mathrm{s}^{-1}; \ 2 - 10^{-1}; \ 3 - 10^{2}; \ 4 - 10^{-1}$.

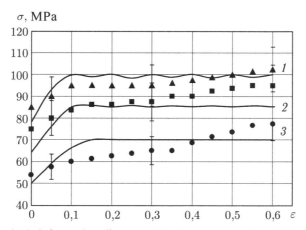

Fig. 7.2. Steel 10 deformation diagrams 10 at temperature $T = 1123$ K (850°C) and strain rates $\dot{\varepsilon}: 1 - 10^{-1}\,\mathrm{s}^{-1}; \ 2 - 10^{-2}; \ 3 - 10^{-3}$.

determined value) is observed in low-carbon steels at $T = 450°C$ in the strain range $\varepsilon \leq 0.3$ (Fig. 7.1, curves 2 and 3). This can be explained by the known non-monotonicity of the dependence of the deformation resistance of these steels on the degree of deformation in the temperature range (300–500)°C, caused by strain ageing – formation of atoms of interstitials (carbon, nitrogen) around the moving and multiplying during plastic deformation dislocations (Cottrell clouds), which increase the flow stress in the range of small strains [70]. For $\varepsilon > 0.2$–0.3, the dislocations break away from impurity segregations. The physico-mathematical theory does not take

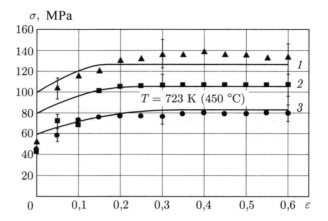

Fig. 7.3. The deformation diagram of 30G1P steel at a temperature of $T = 1123$ K (850°C) and strain rates $\dot{\varepsilon}$: $1-10^{-1}\text{s}^{-1}$; $2-10^{-2}$; $3-10^{-3}$.

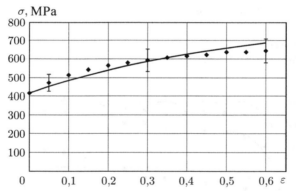

Fig. 7.4. The deformation diagram of 20G2P steel at a temperature of $T = 293$ K (20°C) and a strain rate $\dot{\varepsilon}:10^{-2}\text{s}^{-1}$.

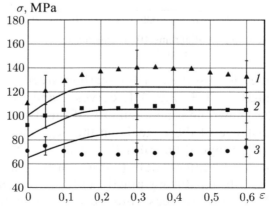

Fig. 7.5. Deformation diagrams of 20G2P steel at temperature $T = 1123$ K (850°C) and strain rates $\dot{\varepsilon}:1-10^{-1}\text{s}^{-1}$; $2-10^{-2}$; $3-10^{-3}$.

into account the temperature intervals of the anomalous behaviour of σ. At these temperatures, processing of these steels with pressure is not carried out.

2. In the remaining investigated cases, the discrepancy between theoretical deformation diagrams and experimental ones does not exceed (10–15)%. In this case, the coefficient β in the expression for the self-diffusion activation energy $U = \beta G b^3$ increases with increasing temperature in accordance with the formula

$$\beta(T) = 5 \cdot 10^{-5} \cdot T + 0{,}3639, \qquad (7.1)$$

where T – is the temperature in K.

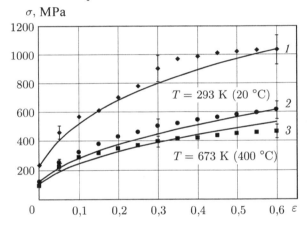

Fig. 7.6. Deformation diagrams of 12Cr18Ni9 steel at various temperatures and strain rates $\dot{\varepsilon}$: *1* – 10^{-1}s^{-1}; *2* – 10^{-2}; *3* – $2 \cdot 10^{-1}$.

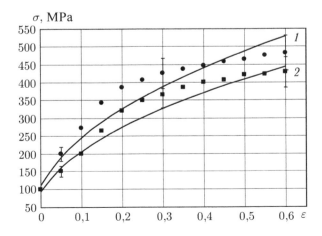

Fig. 7.7. Deformation diagrams of 12KCr18Ni9 steel at a temperature of $T = 873$ K (600°C) and strain rates $\dot{\varepsilon}$: *1* – $3 \cdot 10 \text{ s}^{-1}$; *2* – $2 \cdot 10^{-1}$.

An increase in β with an increase in T indicates an increase in the activation energy U in metals during their deformation under these conditions. This agrees with the known results of the influence of T on U [67]. In deformation under low T conditions tubular (by dislocation cores) and grain-boundary diffusion prevail, and at high T (hot deformation) the main contribution to diffusion mass transfer is made by bulk diffusion (along the body of grains). The activation energy of the latter is higher than that of the pipe and grain boundary diffusion.

In order to verify taking into account the load history model, an experiment was performed with mathematical modelling the uniaxial upsetting of cylindrical samples of steel 10 at $T = 1023$ K (750°C) with a step change $\dot{\varepsilon}$ in the following modes: $\dot{\varepsilon} = $ const $= 10^{-2}\,\mathrm{s}^{-1}$ (curve 1 in Fig. 1.4); to $\varepsilon = 0.4$ with $\dot{\varepsilon} = 10^{-3}\,\mathrm{s}^{-1}$, then with $\dot{\varepsilon} = 10^{-2}\,\mathrm{s}^{-1}$ (curve 2); up to $\varepsilon = 0.4$ with $\dot{\varepsilon} = 1.3 \cdot 10^{-1}\,\mathrm{s}^{-1}$, then with $\dot{\varepsilon} = 10^{-2}\,\mathrm{s}^{-1}$ (curve 3). From Fig. 1.4 it follows that the model describes quite satisfactorily the experimental dependences $\sigma(\varepsilon)$ for different laws of variation $\dot{\varepsilon}(\varepsilon)$ during deformation.

In the work, the adequacy of the model was checked also on pure metals (titanium 99.9%, oxygen-free copper OFHC copper), aluminium alloy AMg and medium carbon steel 45 in the temperature-speed range in which the industry pressure processes these materials. The values of ρ_{so} and λ found for these metals, as well as the values of G for different T, are given in Table 7.2. The values of β and G for oxygen-free copper OFHC are given in Table 7.3. The coefficient

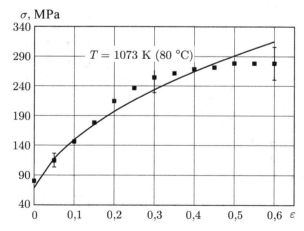

Fig. 7.8. The deformation diagram of steel 12Cr18Ni9 at temperature $T = 1073$ K (800°C) and the strain rate $\dot{\varepsilon} = 10^2\,\mathrm{s}^{-1}$.

Table 7.2. The values of the model parameters used to calculate the deformation diagrams

Material	$\rho_{so} \cdot 10^{10}$, cm^{-2}	$\lambda \cdot 10^{-5}$, cm	\multicolumn{7}{c}{Parameter}
			\multicolumn{7}{c}{$G(T) \cdot 10^3$, MPa}
			\multicolumn{7}{c}{Temperature, °C}
			20	400	480	600	750	800	900
Ti	8.9	6.73	44.9	38.0				31.0	
AMg	1.8	48.3	26.8		30.0				
Steel 45	3.1	9.3	75.5					48.0	45.0

Table 7.3. The values of β and G for OFHC copper at different temperatures

T, °C	25	220	323	457	823
β	0.3788	0.3886	0.3937	0.4004	0.4187
G, MPa	46 000	43 500	44 000	38 500	37 000

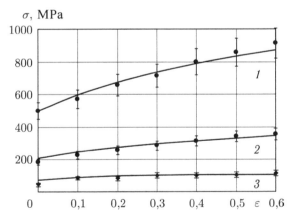

Fig. 7.9. Deformation diagrams of titanium with a strain rate $\dot{\varepsilon} = 2$ s^{-1} at different temperatures (°C): *1* – 20; *2* – 400; *3* – 800.

β was calculated from equation (7.1). At the same time, for copper, $\rho_{so} = 2.36 \cdot 10^9$ cm^{-2}, $\lambda = 1.987 \cdot 10^{-4}$ cm.

The experimental deformation diagrams are borrowed from [83]. The results of the study are shown in Fig. 7.9–7.13.

From the analysis of these results it follows that:

1. The scalar physico-phenomenological model of viscoplasticity with isotropic hardening describes with satisfactory accuracy the deformation diagrams of metals with different types of crystal lattice,

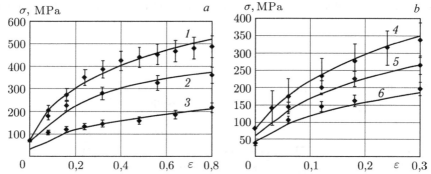

Fig. 7.10. Diagrams of deformation of oxygen-free OFHC copper at $\dot{\varepsilon} = 4 \cdot 10^3$ s^{-1} (*a*) and $\dot{\varepsilon} = 4.5 \cdot 10^2$ s^{-1} (*b*) and temperatures (°C): *1* – 25; *2* – 323; *3* – 823; *4* – 25; *5* – 220; *6* – 457.

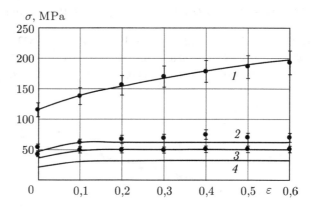

Fig. 7.11. Deformation diagrams of AMg alloy at different temperatures and strain rates: *1* – *T* = 20°C, $\dot{\varepsilon} = 4.0$ s^{-1}; *2, 3, 4* – *T* = 480 °C, $\dot{\varepsilon} = 63, 4, 0.1$ s^{-1}, respectively; *4* – theoretical.

and also alloys based on aluminium and iron at temperatures and rates at which they are subjected to pressure treatment. Titanium has a BCC lattice (low-temperature modification) to 882°C. In this case, the temperature dependence of the coefficient β for it is also described by the equation (7.1).

2. The model describes most accurately the $\sigma(\varepsilon)$ dependences of pure titanium and copper (Figs. 7.9 and 7.8). There is a simple explanation for this. In pure metals, the only barriers to moving dislocations are barriers of a dislocation type. This position formed the basis of the model. In the alloys, other strengthening mechanisms, including solid-solution as well as dispersed particles of other phases, operate. These mechanisms are taken into account in the model

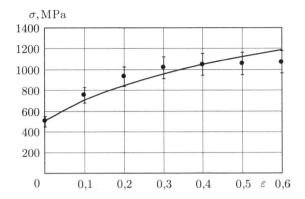

Fig. 7.12. Deformation diagram of steel 45 at temperature $T = 20°C$ and $\dot{\varepsilon} = 5.0$ s^{-1}.

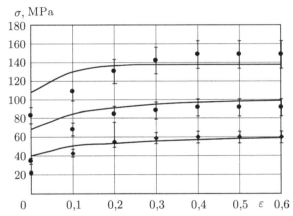

Fig. 7.13. Diagrams of deformation of steel 45 at different temperatures and strain rates: $1 - T = 900$ °C, $\dot{\varepsilon} = 5.0$ s^{-1}; 2 and $3 - T = 1200°C$, $\dot{\varepsilon} = 50$ and 5.0 s^{-1} respectively.

indirectly through parameters ρ_{so} and λ, which are determined for a particular alloy on the basis of a 'cold' deformation diagram.

In order to illustrate the possibilities of the model for describing a wide range of the rheological behaviour of real metals, plastic behaviour of the VT6 titanium alloy (Ti–6.4Al–4.0V–0.16Fe) under the conditions of hot deformation ($T > 0.6T$) was carried out (Fig. 7.14). The calculation was carried out at the following values of the model parameters: $\rho_{so} = 8.96 \cdot 10^{10}$ cm^{-2}; $\lambda = 0.61 \cdot 10^{-4}$ cm; $G = 27$ 700 MPa.

It is established that the model well describes the non-monotonic diagram $\sigma(\varepsilon)$ with the presence of a falling section if it is assumed that the coefficient β depends on the intensity of the accumulated

196 *Physico-Mathematical Theory of High Irreversible Strains*

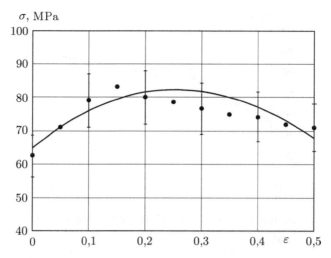

Fig. 7.14. The deformation diagram of the titanium alloy VT6 at $T = 1273$ K ($1000°$C) and $\dot{\varepsilon} = 10$ s^{-1}.

plastic deformation (decreases with increasing ε) and this dependence is described for the VT6 alloy by the equation

$$\beta(\varepsilon) = 0.62 - 9.8 \cdot 10^{-3} \cdot \varepsilon_{(g)}. \tag{7.2}$$

A possible explanation of the dependence $\beta(\varepsilon)$ is as follows. It is known that in the hot deformation of two-phase titanium alloys, even at relatively low degrees of deformation, the process of refining the phase grains is intensively carried out in them. With increasing ε, the grain size decreases. This can lead to a continuous increase in the role of grain-boundary diffusion, the activation energy of which is less than that of bulk diffusion.

7.2. Model of ductile fracture of metals

Some results of the experimental verification of the ductile fracture model are given in [197, 210]. In this paper, in order to verify the fracture model, a plasticity diagram of 20kp steel was experimentally obtained (the points in Fig. 7.15 a). The diagram reflects the dependence of the limiting (before macrofracture, visible with the naked eye) of the shear strain Λ^* from the rigidity index of the stress state scheme $K = \sigma_0/\tau$. The chemical composition of 20kp steel is given in Table 7.4.

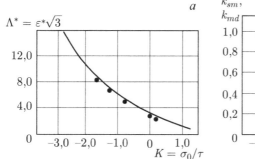 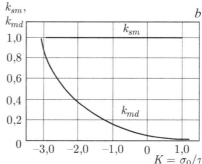

Fig. 7.15. The plasticity diagrams of steel are 20kp (the points are the experiment, the solid curve is theoretical (*a*) and the dependence of the probabilities of formation of k_{sm} and healing k_{md} of microcracks on the rigidity index of the stress state scheme (*b*).

Table 7.4. Chemical composition of steel 20kp

	Chemical composition,% by weight						
Base	C	Si	Mn	Cr	Ni	S	P
Fe	0.17–0.24	0.003–0.07	0.25–0.50	<0.25	<0.25	<0.04	<0.035

The diagram $\Lambda^*(K)$ was constructed using a special technique and using the original setup which are described in [2, 104]. Cylindrical specimens were deformed by torsion (shear) with the simultaneous application of a constant hydrostatic pressure: $p = 0; 250; 500; 700$ MPa. When the specimens were loaded in torsion under constant hydrostatic pressure, the K value changes with increasing strain intensity in accordance with the formula

$$K = -p/\tau(\varepsilon). \tag{7.3}$$

Therefore, each point in Fig. 7.15 *a* characterizes the limiting deformation corresponding to the loading history $K(\varepsilon)$ described by formula (7.3).

It is known that metals at deformation intensities greater than 1–2 behave as ideally plastic bodies, i.e. deformation occurs at $\sigma = $ const [57]. Limiting strains for $K < 0$ are greater than two. Therefore, in constructing the theoretical diagram $\Lambda^*(K)$, the value of $K(g)$ was calculated from (7.3) to $\varepsilon = 2.0$. For $\varepsilon > 2.0$, $\tau_{(g)} = $ const was assumed.

When testing specimens with torsion, the deformation is monotonous and the load is simple. Therefore, at each step g, taking into account $\tau = \sigma/\sqrt{3}$ and $\Lambda = \varepsilon\sqrt{3}$, the calculation was carried out sequentially by the formulas: (4.16); (4.35); (4.36); (4.37);

(4.26); (4.27) and (4.28). The value of $\Lambda_{(g)}$ at which $\psi_{(g)} = 1.0$ was considered equal to $\Lambda^*_{(g)}$ and put in accordance with the value of $K_{(g)}$ at the same calculated step g. The calculations were carried out at the following values of the parameters of the model for 20kp steel: $G = 84\,800$ MPa; $b = 3 \cdot 10^{-8}$ cm; $\lambda = 0.713 \cdot 10^{-4}$ cm; $\bar{\xi} = 0{,}32 \cdot 10^{-5}$ cm; $\rho_{so} = 2.74 \cdot 10^{10}$ cm^{-2}. The dependences $k_{sm}(K)$ and $k_{md}(K)$, calculated in accordance with (4.32) and (4.33), were preliminary calculated in Fig. 7.15 b. It is seen that the probability of healing microcracks for steel 20kp $k_{md} = 1.0$ at $K = -3.1$. Therefore, at $K \leq -3.1$ steel should show practically unlimited plasticity. However, it is technically difficult to show this experimentally due to the high value of $p = -\sigma_0$.

When the viscous fracture model was verified, four different steels were used to determine that the critical density of microcracks N^* depends on the rigidity index of the stress state K.

For 'soft' schemes (when compressive stresses prevail) $K < -2.5$, the critical density $N^* = 10^7$ cm^{-2}. In the case of 'hard' schemes of the stress state $K > 0.58$ (tensile stress prevails), $N^* = 10^6$ cm^{-2}. For intermediate values of $K \in [-2.5;\, 0.58]$, the critical density of microcracks depends on K and is described by equation

$$N^*(K) = -60.2532 \cdot 10^4 K^3 - 3 \cdot 10^6 \cdot K^2 + 8 \cdot 10^6 \left[\text{cm}^{-2}\right]. \qquad (7.4)$$

This result has a simple physical explanation. At high positive hydrostatic pressure values, combining the ensemble of microcracks into the macrocrack is difficult, and with negative hydrostatic pressure, the opposite happens faster.

From Fig. 7.15 a it follows that the physico-phenomenological model of ductile fracture describes quite satisfactorily the experimental results. This allows us to recommend a model for use in mathematical modelling of technological processes of metal working with pressure.

7.3. Creep model

The creep curves of alloys based on Fe and Ni were calculated using the equations given in Section 6.2, which were compared with the experimental ones. The initial values of ρ_{s0} and λ were determined from the 'cold deformation diagrams', as described in Section 4.3.

The calculation program included a comparison of the dislocation densities at the neighboring calculation steps g and $(g + 1)$. If it the condition $\Delta\rho = \rho_{(g+1)} - \rho_{(g)} \leq 10^6$ cm^{-2} was fulfilled, then, at an average initial dislocation density in alloys of the order of $\rho_{s0} = 10^9$–10^{10} cm^{-2}, it was assumed that $\Delta\rho \approx 0$ and the stage of steady creep for which $\partial\rho/\partial t = 0$ started.

The value of $\rho_{(g+1)}$, at which $\Delta\rho \leq 10^6$ cm^{-2}, was taken as the steady-state ρ^c. Using the structural law of steady-state creep (6.11), the corresponding ρ^c value of the mean free path of dislocations was determined as

$$\lambda^c = 1.0/\rho^c b. \tag{7.5}$$

In accordance with (6.13), the creep strain rate at the steady state with allowance for (7.5) will be

$$\dot{\varepsilon}^c_{min} = (\rho^c)^{1/2} v_D b \exp\left(-\frac{\beta(T,\sigma)G(T)b^3 - \sigma^c b^2/m\sqrt{\rho^c}}{kT}\right). \tag{7.6}$$

The creep strain at the steady state was calculated as $\varepsilon^c = \dot{\varepsilon}^c_{min} \cdot t$.

The algorithm for verifying the creep model provided verification of the validity of the structural creep law (7.5), and the equations for the minimum velocity (7.6).

As an example, Figs. 7.16, 7.17 and 7.18 show the experimental and theoretical creep curves for EI696, 2Kh13 and EI826 steels, respectively. The segments at the experimental points show a symmetrical ten percent deviation from the mean experimentally determined value of ε^c. The experimental dependences are taken from [218, 219].

The chemical composition of the materials is given in Tables 7.5 and 7.6.

When calculating the creep curves $\varepsilon^c(t)$, the values of the model constants were as follows: $b = 3.1 \cdot 10^{-8}$ cm; $m = 3.1$; $k = 1.38 \cdot 10^{-23}$ J · deg^{-1}; $v_D = \dot{\varepsilon}_* = 10^{12}$ s^{-1}.

The values of the shear moduli adopted at the calculation for different temperatures are taken from [209] and are given in Table 7.7.

The dependences of the coefficient β determined from the experimental creep curves in the formula for the activation energy of self-diffusion ($U = \beta G b^3$) on temperature and stress are given in Table 7.8.

The values of the mean free path of dislocations λ and the initial dislocation density ρ_{s0} in the material calculated from the 'cold' deformation diagrams $\sigma(\varepsilon)$, according to the procedure described in Section 4.3, have the following values: steel EI696 – $\lambda = 2.37 \cdot 10^{-5}$ cm, $\rho_{s0} = 5.66 \cdot 10^{10}$ cm^{-2}; alloy EI826 – $\lambda = 3.32 \cdot 10^{-5}$ cm, $\rho_{s0} = 6.83 \cdot 10^{10}$ cm^{-2}; steel 2Kh13 – $\lambda = 2.43 \cdot 10^{-4}$ cm, $\rho_{s0} = 3.94 \cdot 10^{10}$ cm^{-2}.

As another example Fig. 7.19 shows the creep curves for steel 08 at a constant stress $\sigma = 100$ MPa and various temperatures.

200 *Physico-Mathematical Theory of High Irreversible Strains*

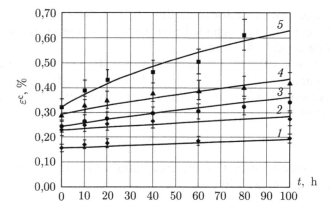

Fig. 7.16. Creep curves of EI696 steel at a temperature of 700°C and stresses: *1* – 170, *2* – 230, *3* – 300, *4* – 320, *5* – 360 MPa (points – experiment, solid curves – theory).

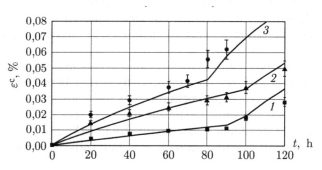

Fig. 7.17. Creep curves of steel 2Kh13 at a temperature of 600°C and stresses: *1* – 90 MPa, jump to 120 MPa, *2* – 110 MPa, jump to 130 MPa, *3* – 140 MPa (points – experiment, line – theory).

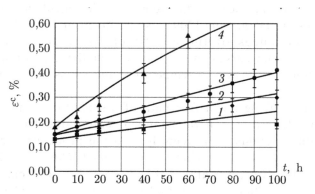

Fig. 7.18. Creep curves of the EI826 alloy at 800°C and stresses: *1* – 200, *2* – 250, *3* – 260, *4* – 280 MPa (points – experiment, solid curves – theory).

Experimental Verification of Adequacy of Models 201

Table 7.5. The chemical composition of EI696 and 2Kh13 steels in%

	Fe	C	Si	Mn	Ni	S	P	Cr	Ti	Al	B
EI696	basis	to 0.1	to 1.0	to 1.0	18–21	to 0.02	to 0.035	10–12.5	2.6–3.2	to 0.8	to 0.02
2Kh13	basis	0.16–0.25	to 0.6	to 0.6	to 0.6	to 0.025	to 0.03	12–14	—	—	—

Table 7.6. The chemical composition of EI696 and 2Kh13 steels in%

Ni	Fe	C	Si	Mn	S	P	Cr	Ce	Mo	W	V	Ti	Al	B
75,2 60.621– basis	to 5	to 0.12	to 0.6	to 0.5	to 0.009	to 0.015	13–16	to 0.02	2.5–4.0	5.0–7.0	0.2–1.0	1.7–2.2	2.4–2.9	to 0.015

Table 7.7. The values of G of materials for different T

EI 696 steel		EI 826 alloy		Steel 2Kh13	
T, °C	G, MPa	T, °C	G, MPa	T, °C	G, MPa
20	666 666	20	74 074	20	—
700	41 233	800	53 704	600	63 000

Table 7.8. The values of parameter β for materials at different creep stresses

Steel EI696 ($T = 700°C$)		Alloy EI826 ($T = 800°C$)		Steel 2Kh13 ($T = 600°C$)	
σ^c, MPa	β	σ^c, MPa	β	σ^c, MPa	β
170	0.710	200	0.585	90	0.357
230	0.770	250	0.619	110	0.365
300	0.836	260	0.622	120	0.368
320	0.855	280	0.628	130	0.375
360	0.885	—	—	140	0.381

The chemical composition of the steel is given in Table 7.9. The calculation was made using formulas (6.3)–(6.8) with the following values of the model parameters: $\lambda = 3.93 \cdot 10^{-4}$ cm; $\rho_{so} = 5.25 \cdot 10^9$ cm^{-2}; $dt_{(g)} = 300$ s. The values of G and β at different temperatures, which are accepted for calculation, are given in Table 7.10.

The experimental creep curves for steel 08 were taken from [211].

It follows from the above results that the model developed well describes the experimental dependences $\varepsilon^c(t)$, including those obtained under conditions of unsteady loading. In this case, the activation energy of creep (the coefficient β in the expression $U = \beta G b^3$) depends on temperature and stress. With increasing T at $\sigma^c =$ const, the coefficient β increases. With increasing σ^c at $T =$ const β also increases.

Since the coefficient β has a specific physical content in the model – it characterizes the activation energy of the elementary act of the process of thermally activated slip of dislocations, which has a diffusion nature, the dependences $\beta(T)$ and $\beta(\sigma)$ can have the following explanation.

In creep under low T conditions, pipe (by dislocation cores) and grain-boundary diffusion prevail, and at high T the main contribution to diffusion mass transfer is made by bulk diffusion (along the body

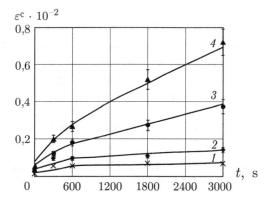

Fig. 7.19. The creep curves for steel 08 at a stress $\sigma = 100$ MPa and temperatures: 1 – 400; 2 – 425; 3 – 450; 4 – 460°C (points – experiment, solid curves – theory).

Table 7.9. Chemical composition of steel 08

Chemical element	C	Mn	Si	Cr	S	Cu	As	P	Ni
								no more	
Content,%	0.05–0.12	0.35–0.65	0.17–0.37	0.1	0.04	0.25	0.08	0.035	0.25

Table 7.10. The values of G and β at different temperatures for steel 08

Temperature, °C	400	425	450	460
G, MPa	60 000	59 000	58 000	57 500
β		0.462	0.472	0.477	0.478

of grains). The activation energy of the latter is higher than that of the pipe and grain boundary diffusion [68].

It is also known [44] that with increasing σ^e the share of diffusion fluxes along the grain body increases, which also leads to an increase in the activation energy, i.e., to an increase in β.

7.4. Stress relaxation model

The stress relaxation curve in steel 30KhM was calculated from the formulas (6.38)–(6.43) for the following values of the constants and model parameters at $T = 500°C$: $E = 176\,600$ MPa, $G = 66\,415$ MPa, $\rho_{so} = 8 \cdot 10^{10}$ cm^{-2}; $\lambda = 4.16 \cdot 10^{-4}$ cm, initial stress $\sigma_{(1)} = 200$ MPa, $k = 1.38 \cdot 10^{-23}$ J/K, $dt_{(g)} = 10$ hours $= 36\,000$ sec, $\beta = 0.402$.

The chemical composition of 30KhM steel is shown in Table 7.11. The experimental dependence $\sigma(t)$ was taken from [7].

The theoretical and experimental stress relaxation curves are shown in Fig. 7.20. It can be seen that the stress relaxation model satisfactorily describes the experimental dependence $\sigma(t)$. Apparently, a more accurate coincidence of the experimental and theoretical curves will be obtained by taking into account the dependence of β on stress.

7.5. Model of long-term strength

Verification of the physico-phenomenological model of long-term strength was carried out by comparing the results of mathematical modelling of the process of testing samples for long-term strength under the conditions of stationary thermomechanical loading with experimental data. The time to fracture was determined before the sample of steel 20 was fractured at a temperature of 500°C and uniaxial tension. The values of the parameters and constants of the model are given in Table 7.12.

The time to fracture of the sample t^* for different σ was calculated using the formulas (6.34)–(6.36).

Table 7.11. Chemical composition of steel 30KhM

Chemical element	C	Si	Mn	Cr	Mo	P	S	Cu	Ni
						\multicolumn{4}{c}{no more}
Content, %	0.26–0.33	0.17–0.37	0.4–0.7	0.8–1.1	0.15–0.25	0.035	0.035	0.3	0.3

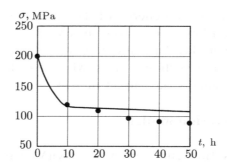

Fig. 7.20. The stress relaxation curves in 30KhM steel at $T = 500°C$ and an initial stress of 200 MPa (the points are the experiment, the solid curve is the theory).

Table 7.12. Values of constants and parameters for steel 20

$\bar{\xi}$, cm	ν_D, s^{-1}	b, cm	$G^{500°C}$ MPa	$m = M$	kT, J	N_m^*, cm^{-2}	λ, cm	ξ_0	K
$0.32 \times \times 10^{-5}$	10^{12}	$3 \cdot 10^{-8}$	65 000	3.1	$1066.74 \cdot 10^{-23}$	10^6	$0.73 \cdot 10^{-4}$	$0.32 \cdot 10^{-5}$	0,58

The parameter β for different stresses has the following values: $\sigma_{long} = 200$ MPa $- \beta = 0.43$; $\sigma_{long} = 170$ MPa $- \beta = 0.4$; $\sigma_{long} = 110$ MPa $- \beta = 0.34$.

The results of the simulation are shown in Fig. 7.21. The experimental dependences $\sigma_{long}(t^*)$ are taken from [7].

The results testify – the physico-phenomenological model of long-term strength can be used for design calculations.

The concrete results of calculating the dynamics of accumulation and healing of deformation microcracks in a sample under the conditions of uniaxial tension ($K = \sigma_0/\tau = 0.58$) at $T = 500°C$ and three different stresses, which are presented in Table 7.13, are of independent interest.

In the table, $N_{m\ samp}$ is the theoretical value of the scalar density of microcracks formed in the sample during the time t^* and at a stress σ_{long} at which the sample breaks down and which are established experimentally; $N_{m\ heal}$ are respectively the scalar density of microcracks, healed under the same conditions; N_m^* is the experimentally established critical density of microcracks under uniaxial tension.

It is seen that under the conditions of tensile stresses, when the rigidity index of the stress state K is positive and equal to 0.58, microcracks do not heal. This is explained by the high activation

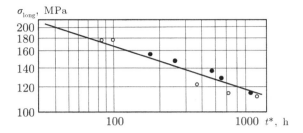

Fig. 7.21. Dependence of fracture stress intensities on time for steel 20 (points – experiment, solid line – theory). Black points – testing of samples under uniaxial tension; light points – testing of thin-walled tubes loaded with internal pressure.

Table 7.13. The results of calculating the long-term strength by formulas (6.30)

N_m, cm^{-2}	$N_{m\,heal}$, cm^{-2}	σ_{long}, MPa	t*, h	N_m^*, cm^{-2}
$1.237 \cdot 10^6$	0	110	1000	10^6
$1.135 \cdot 10^6$	0	170	100	
$1.268 \cdot 10^6$	0	200	30	

energy of this process under the given deformation conditions (the numerator of the exponential expression in the second term of the equation (6.30) Evidently, another experimentally established fact [127], which is theoretically established here, is explained, namely, that the known fracture criteria that do not take into account the healing of microcracks, provide a more or less satisfactory prediction of failure at $K > 0$. For $K < 0$ they do not give a prediction of failure corresponding to the experiment.

7.6. Model of evolution of the structure in processes of irreversible deformation of metals

The theory presented in Section II is constructed using the structural-phenomenological approach and contains the evolution equations for the scalar dislocation densities, microcracks and average linear grain size and the subgrain functionals $\rho[\varepsilon(t),\dot{\varepsilon}(t),T(t),t]$, $N_m[\varepsilon(t),\dot{\varepsilon}(t),T(t),K(t),t]$ and $D[\varepsilon(t),\dot{\varepsilon}(t),T(t),\rho(t),t]$. This allows, firstly, to take into account the influence of these continuously changing structural characteristics on the mechanical state of the deformable metal, and secondly, to predict the properties and quality of the metal after deformation, which is especially important in the pressure working of metals and design calculations for creep and long-term strength.

The scalar density of dislocations, for example, in the deformed workpiece, which is deformed by methods of metal working, at any time and in any elementary volume, can be estimated using the equations (4.12), (4.16), (4.22) and (6.2). Knowing the scalar dislocation density, we can estimate the average linear size of grains, fragments or subgrains from equation (6.1). This is important for the design of processes for the plastic structuring of metals [34, 154, 94].

To illustrate what has been said, Fig. 7.22 shows the experimental (points) and calculated by (4.23) (solid curve) deformation diagram $\sigma(\varepsilon)$ of the AMg aluminium alloy at $T = 20°C$, and the dependences of the scalar dislocation density ρ and the linear the size of subgrains D on the intensity of deformation. In Fig. 7.22 the values of $\rho \cdot 10^{10}$ [cm^{-2}] and $D \cdot 10^{-5}$ [cm] are plotted on the right axis of the ordinates. The calculations are performed up to large values of ε.

It can be seen that all three dependences reach saturation at $\varepsilon = 3.0$. The theory predicts a decrease in the grain size with an increase in the degree of cold deformation. The technology of plastic structure formation of metals is based on this phenomenon [34, 94, 154]. In particular, the developed theory predicts the possibility of obtaining a microcrystalline state of the AMg aluminium alloy having a grain size of 0.4 μm by cold deformation with an intensity $\varepsilon = 3.0$. In this case, the alloy will have a yield strength $\sigma_T \cong 260$ MPa (Fig. 7.22).

As a specific example Fig. 7.23 shows the experimental (points) and theoretical (solid curves) of the dependence of the average linear dimension of the fragments D on the accumulated intensity of strain ε for aluminium Al (99.99%) and Al + 3% Mg alloy. The experimental dependences are taken from [212]. In the experiments, the samples were deformed at $T = 20°C$ by the method of equal-channel angular pressing (ECAP process) with different number of passes.

Between the passes the samples were not turned over. Therefore, in the set of passes, the loading can be considered simple, and the deformation is monotonic. In this case, the theory gives the equation for the dislocation density in the finite form (4.22).

From Fig. 7.23 it follows that the model with good accuracy allows us to predict the linear size of fragments. For these materials, $B = 2.0$. Calculations were carried out for the following values parameters of the model: Al (99.99%) $\rho_{so} = 0.254 \cdot 10^{10}$ cm^{-2}, $\lambda =$

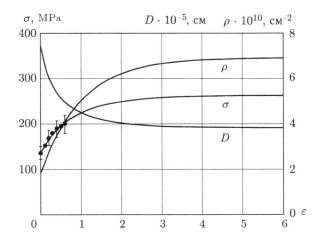

Fig. 7.22. The dependence of the stress intensity σ, the scalar dislocation density ρ and the average linear grain size D on the intensity of the accumulated plastic deformation in the AMr alloy.

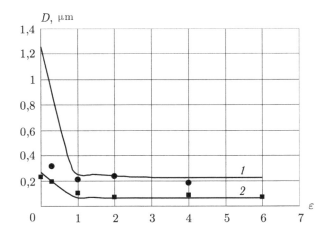

Fig. 7.23. Dependences of the average linear size of fragments on the accumulated intensity of plastic deformation: 1 – Al (99.99%); 2–Al + 3% Mg.

$409.0 \cdot 10^{-5}$ cm, $G = 28\,000$ MPa; Al +3% Mg $\rho_{so} = 5.58 \cdot 10^{10}$ cm^{-2}, $\lambda = 31.4 \cdot 10^{-5}$ cm, $G = 27\,000$ MPa.

The difference B from 10.0 for these materials may be due to the fact that the size of the fragments could be determined experimentally at sample sites that underwent the most intense deformation, for example, in contact layers. Perhaps the average size of fragments by volume is somewhat larger.

208 *Physico-Mathematical Theory of High Irreversible Strains*

The values of ρ_{so} and λ in all the calculations given in this paper were determined by the method described in section 4.3.

7.7. The model of a large cyclic and near-plastic deformation

To solve the problem of experimental verification of the developed plasticity model, a process of cold volume stamping (CVS) of M24 × 70 bolts from steel 20 was chosen [213].

The stamping processes are shown in Fig. 7.24. The initial hot-rolled bar of diameter 25 mm after cleaning the surface by etching and applying a phosphate coating is calibrated by drawing to a diameter of 23.5 mm (Fig. 7.24 a). Then, stamping is carried out on cold-setting machines. Forming includes: pre-insertion of the head, final upsetting of the head, reduction of the rad for threading to form a chamfer and trimming the hexagon (Fig. 7.24, b–e, respectively).

The experimental dependences of the deformation force (P) – the displacement of the punch (S) during the operation of dislodging the head of the bolt (the first and second forming transitions (Figs. 7.24 b and c) are given in [213].) Therefore, model verification was compared by comparing the experimental and theoretical P(S) dependences. The latter were mathematically modelled in the shaping processes in the designated cold volume stamping transitions on the basis of the developed plasticity model.

In accordance with the technology of material preparation for cold forging, high-quality steels are subjected to spheroidizing annealing

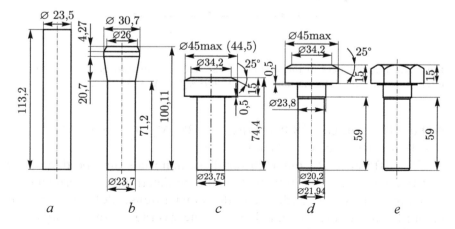

Fig. 7.24. Technological transitions of cold volume stamping of M24 × 70 bolts.

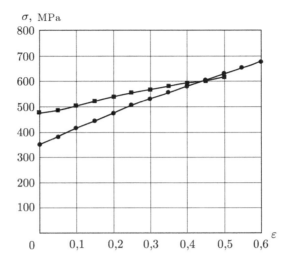

Fig. 7.25. Experimental diagrams of deformation of steel 20: ● – initial state (after spheroidizing annealing); ■ – upsetting after preliminary deformation according to the scheme of simple stretching to $\varepsilon^+ = 0.51$ (rod drawing).

to increase ductility and reduce the deformation resistance, i.e., in the initial state, the steel is isotropic.

According to the procedure described in 5.2, using the experimental deformation diagrams of steel 20, which are shown in Fig. 7.25, the parameters of the plasticity model of a large cyclic and cold deformation close to it were determined. The values of the parameters used fo the calculations are given in Table 7.14.

The stress function $\Phi_0(\varepsilon)$ calculated for equation (5.5) for steel 20 is shown in Fig. 7.26.

The mathematical model of the physical process of metal deformation in the first and second transitions included: a) the equilibrium equations (the volume forces of gravity and inertia were neglected because of their smallness in comparison with the deformation forces)

$$\frac{\partial(d\sigma_{ij})}{\partial x_j} = 0, \quad i,j = x,y,z; \tag{7.7}$$

b) the geometric Cauchy relations

$$d\varepsilon_{ij} = \frac{1}{2}\left(\frac{\partial(du_i)}{\partial x_j} + \frac{\partial(du_j)}{\partial x_i}\right), \tag{7.8}$$

Table 7.14. The values of the model parameters (51) for steel 20

Material	$\rho_{so} \cdot 10^{10}$, cm^{-2}	λ_c, 10^{-4} cm	$A \cdot 10^{10}$, cm^{-2}	σ_{Tcomp}^{exp} MPa	$\sigma_{T,}$ MPa	ε^+	β	G, MPa
Steel 20	1.23	5.52	2.02	475	300	0.51	0.4	85 000

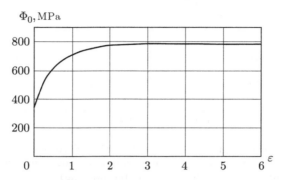

Fig. 7.26. The theoretical function of the stress $\Phi_0(\varepsilon)$ for steel 20.

where d_{ui} are the projections to the coordinate axes of the vector of the increment of the displacement of the point; c) determining relations of the physico-phenomenological model of plasticity for multi-transient deformation processes with non-monotonic large deformation and complex loading in the set of stamping transitions (5.1) and (5.5).

Calculation schemes for the numerical mathematical modelling of deformation are shown in Fig. 7.27.

On the contact surface of the deformable workpiece with the punch, the kinematic boundary conditions were set in the form of equal displacements of the boundary node points d_{uy}, on contact with the matrix, static boundary conditions in the form of forces ensuring the continuity of the contact nodes from the surface. Contact friction was taken into account according to Siebel, $\tau_k = \mu\sigma$, where τ_k are the tangential contact stresses, σ is the stress intensity, and μ is the friction factor. For the cold volum sptamping process, $\mu = 0.12$ is recommended.

The discretization of the continuous mathematical model (7.6), (7.7), (5.1), (5.5) by the finite element method, the realization of the variational principle, allowance for boundary conditions, the solution of a system of algebraic equations, deformation and deformation forces were determined in the DEFORM-3D environment. The

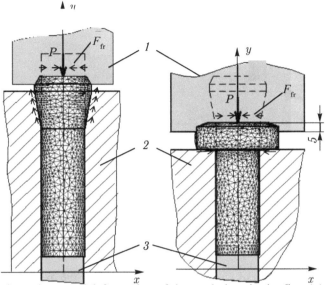

Fig. 7.27. Calculation schemes of deformation of the workpiece in the first (*a*) and second (*b*) punching transitions: *1* – punch, *2* – matrix; *3* – ejector.

deformed volume was divided into 20 000 finite elements in the form of tetrahedra [214, 215].

The software product DEFORM-3D carries out the transition to a discrete finite element model with the determination of the true displacements of the node points on the basis of the Lagrange potential energy minimum variational principle.

The system of differential equations of problem (7.6), (7.7), (5.1), (5.5) is replaced by a system of linear algebraic equations of the form

$$\mathbf{F} = \mathbf{KU}, \qquad (7.9)$$

where **F** is the matrix vector of nodal forces; **K** is the system stiffness matrix; **U** is the matrix vector of nodal displacements.

On the displacements found with allowance for boundary conditions, there are strains, on them – stresses and on the latter – forces.

As a result of mathematical modelling of two technological transitions, fields of accumulated intensity of deformation were obtained (see Figs. 7.28 and 7.29)[1] and the deformation force of the

---

1) Taking into account the symmetry of the detail, the fields in the meridian section are shown.

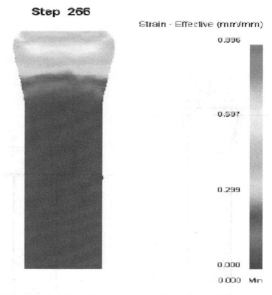

Fig. 7.28. Fields of cumulated strain intensity in the axial section of the blank at the end of the first stamping pass.

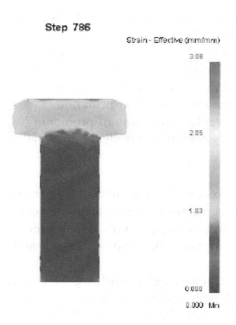

Fig. 7.29. Fields of cumulated strain intensity in the axial section of the blank at the end of the second stamping pass.

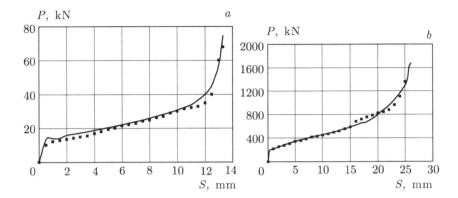

Fig. 7.30. Experimental (points) and theoretical (solid curves) dependences of the deformation force on the displacement of the punch: *a* – preliminary upsetting of the head; *b* – final upsetting of the head.

workpiece in the die as a function of the displacement of the punch (Figs. 7.30, *a*, *b*).

The largest accumulated strains take place in the head of the bolt. At the first upset pass, the average volume of the processed part the value of the accumulated strains is $\varepsilon = 0.5$ (Fig. 7.28), and at the second transition it grows to $\varepsilon = 1.4$ (Fig. 7.29). In the region of transition from the head to the rod, a large strain gradient is observed, since the rod part is practically not deformed.

The maximum calculated specific deformation forces were determined by the formula $p = P/F$, where P is the punching force at the time of completion of the metal filling of the die cavity, F is the area of the projection of the workpiece to the horizontal plane.

At the first pass, $p = \dfrac{680 \text{ kN}}{739.85 \text{ mm}^2} = 919$ MPa, at the second – $p = \dfrac{1870 \text{ kN}}{1589.625 \text{ mm}^2} = 1176.4$ MPa ·

The specific punching forces do not exceed the permissible load on the tool, which, in cold stamping, is $p < 2000$ MPa. At $p \geq 2000$ MPa, the tool life decreases dramatically. Consequently, it is possible to commercialize this bolting technique.

From Fig. 7.30 (*a* and *b*) it follows that the theoretical $P(S)$ dependences almost ideally coincide with the experimental ones at $\mu = 0.12$.

The results of mathematical modeling with the use of a structural-phenomenological model of material plasticity under cyclic and close to it deformation with large strains in cycles make it possible to conclude that the model adequately describes the formation in cold stamping processes characterized by complex loading and large accumulated strains. Therefore, the developed model of plasticity can be recommended for the certification of existing in engineering and the development of new multi-transient cold volume stamping processes with application of CAE-technologies.

8

Mathematical formulation and examples of solving applied problems of the physico–mathematical theory of plasticity

8.1. Mathematical formulation of problems

A unified physico-mathematical theory of irreversible strains and ductile fracture of metals makes it possible to solve the following applied problems [216].

1. In the processes of metal forming (rolling, pressing, drawing, forging, stamping) under conditions of cold, warm and hot deformation, determine the characteristics of the stress–strain state, the deformation force, the evolution of the structure with the determination of the dislocation density and the estimate of the linear dimensions of the substructure, damage to the metal and evaluate the strength characteristics of the metal after deformation. In this case, it is possible to correctly analyze non-stationary deformation processes with complex loading and non-monotonic deformation, as well as deformation processes of metals with non-monotonic deformation diagrams.

The information obtained allows one to choose the main equipment in terms of capacity and type, design tools, rationalize processes for the parameters of interest: energy intensity, cost, quality of metal products, and others.

2. Determine the distribution of stresses and strains in structures and parts, as well as the life of structures and parts by the permissible technical conditions of creep strain or creep rate, long-term strength

under loading by constant forces, and also under the conditions of unsteady loading.

3. Determine the service life of structures with a previously created stress state and with fixed strains operating at elevated temperatures. For example, cylinders and pipes assembled with interference, high pressure vessels, etc.

4. To theoretically analyze various variants of technological processes of plastic structure formation of metals with an assessment of the possible production of the linear dimensions of the substructure and the strength and plasticity characteristics of the resulting microcrystalline metal.

It is known that the solution of these problems in the mechanics of a deformed solid reduces to analyzing the corresponding physical fields in a particular spacetime region in which the process under investigation occurs, that is, solutions of the corresponding initial boundary-value problems are analyzed. The mathematical formulation of boundary-value problems contains the equations of motion of a continuous medium (or the equilibrium equation in the case of neglecting body forces) (7.6) and the Cauchy kinematic relations (7.7) or, in the formulation of the problem in velocities – St. Venant:

$$d\dot\varepsilon_{ij(g)} = \frac{1}{2}\left(\frac{\partial(dv_{i(g)})}{\partial x_{j(g)}} + \frac{\partial(dv_{j(g)})}{\partial x_{i(g)}}\right), \qquad (7.9)$$

where dv_i are the projections onto the coordinate axes x_i of the particle velocity vector.

The system of differential equations (7.6), (7.7) is closed by the equations of the model of a deformable medium – the defining relations (4.46) or (5.1).

When the problem is formulated at velocities, for example for the analysis of creep, equation (4.46) by differentiation with respect to time is reduced to the form

$$d\dot\varepsilon_{ij(g)} = \frac{3}{2}\left(\frac{d\dot\varepsilon_{(g)}}{\sigma^T_{(g)} + d\sigma^u_{(g)} - d\sigma^r_{(g)}}\left(s^T_{ij(g)} + ds^u_{ij(g)} - ds^r_{ij(g)}\right)\right). \qquad (7.10)$$

In the expressions for $d\sigma^u_{(g)}$ and $d\sigma^r_{(g)}$, (4.18) and (4.19), we use the substitution $d\varepsilon_{(g)} = \dot\varepsilon_{(g)} dt_{(g)}$

The above system of equations is a mathematical model of internal mechanisms that occur in the volume of a deformed body. It does not take into account its interaction with the environment. Therefore, the system of equations is supplemented by the boundary conditions (initial and boundary), which constitute the condition of uniqueness of the solution of the particular problem under consideration.

The peculiarity of the mathematical formulation of the boundary value problems of the physico-mathematical theory of irreversible strains of metals is their formulation in increments (see equations (7.5), (7.6), (7.8) and (7.9), which, firstly, in analyzing large irreversible strains, of the non-linear tensor of large Green strains, the Cauchy tensor of small strains (7.7) Secondly, to carry out the summation of strains, stresses, dislocation densities and microcracks along the particle deformation path. Thirdly, in the numerical solution of the system of equations by the finite element method, the computational error is reduced, since at each small step of loading g in each element a quasi-elastic problem with an instantaneous secant modulus is solved (analog of the modulus elasticity E) $E^*_{(g)}$, which is calculated from the scalar plasticity model as

$$E^*_{(g)} = d\sigma^u_{(g)}/d\varepsilon_{(g)}, \qquad (7.11)$$

where $d\sigma^u_{(g)}$ has the form (4.18).

In order to formulate the equations of a quasi-elastic problem in a matrix form and to formulate a quasielastic matrix, which is necessary for the finite element method, the analogues of the shear modulus $G^*_{(g)}$, the Poisson covariance coefficient $v^*_{(g)}$ and the volume compression quasimodule $K^*_{(g)}$ are defined as

$$G^*_{(g)} = E^*_{(g)}\Big/\big[2\big(1+v^*_{(g)}\big)\big], \qquad (7.12)$$

$$K^*_{(g)} = E^*\Big/\big[3\big(1-2v^*_{(g)}\big)\big], \qquad (7.13)$$

$$v^*_{(g)} = 1/2 - (1-2v)E^*_{(g)}/(2E), \qquad (7.14)$$

where v and E are the elastic constants of the material (Poisson's ratio and longitudinal modulus of elasticity).

The system of differential equations of the problem is always closed by the first equation (4.46) (when it is expanded into three equations), as indicated in section 4.5. The deformation diagram $\sigma^u(\varepsilon)$ describing the hardening component always satisfies the deformation stability condition $d\sigma^u/d\varepsilon > 0$.

The solution of the problem is the fields $d\varepsilon_{ij}(x, y, z, t)$ and $\sigma_{ij}^u(x,y,z,t)$. By setting $d\varepsilon_{ij}, \sigma_{ij}^T(x,y,z,t)$ is found in the second equation (4.46), and using of the third equation (4.46) is determined by $\sigma_{ij}^r(x,y,z,t)$. The resulting stress is the summation of its components.

8.2. Examples of development, research and improvement of processes of processing of metals by pressure on the basis of mathematical modelling

The technology of cold volumetric stamping (CVS) is one of the promising technologies of metal forming. CVS provides a plastic molding of finished parts that go to the assembly. At the same time, the metal utilization ratio is the highest among other metal processing technologies and is 98%. We manufacture parts with high dimensional accuracy, high surface quality and high productivity (50–300) pieces per minute. Typically, the technology is used in large-scale and mass production. Sutomated equipment and systems are used. Therefore, any improvement in technology in terms of energy intensity, tool life, quality of metal products brings a significant economic effect.

In order to demonstrate the application of the new theory to the improvement and development of CVS processes, let us consider two examples.

To increase the productivity and the coefficient of metal utilization by replacing the technology of shaping by cutting with the technology of plastic shaping on the cold-setting machine, the experimental process of the CVS of a 'bushing' component of steel 20 was developed on the basis of available technological recommendations and production experience. The stages of the experimental process are shown in Fig. 8.1.

The task of certification of technology by the method of mathematical modelling was made on the basis of the physical and mathematical theory of plasticity. Stamping is carried out at room temperature (preheating in deformation to 150–200°C is neglected). It follows from Fig. 8.1 that the deformation of the billet in the set of transitions takes place under conditions of complex loading and non-monotonic deformation. Therefore, the mathematical formulation of the problem included the equations of the physico-mathematical model of cyclic and plastic deformation, close to it, i.e., the problem was formulated and solved in a manner analogous to that described in section 7.7. The difference was, first, that in accordance with the geometry of the workpiece and the loading scheme of the blanks

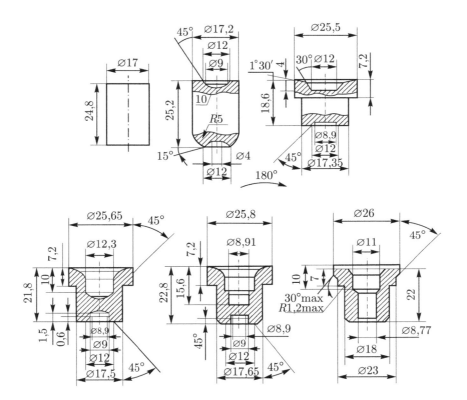

Fig. 8.1. Transitions of the experimental bushing stamping process.

in the dies, the problem was posed and solved in the DEFORM-3D environment as an axisymmetric problem.

Secondly, in parallel with the calculation of the characteristics of the stress–strain state, the distribution of the density of microcracks in the deformed volume and the probability of macrofracture $\psi(g)$ by the criteria (4.27) and (4.28) were determined. For this purpose, an additional calculation program was compiled using the equations (4.25) and (4.26).

As a result of shaping simulation in punching stages, it was established that starting from the second punching stage the model predicts the formation of a macrocrack in different parts of the billet, i.e. in these places $\psi(g) = 1.0$. In Figure 8.2 these places are shown by arrows.

Analysis of simulation results showed that the places of possible occurrence of macrocracks ($\psi = 1.0$) coincide with the places of the greatest intensity of accumulated strains, which have values from 2.6 after the completion of the second transition to 3.6–4.0 at the fifth

220 *Physico-Mathematical Theory of High Irreversible Strains*

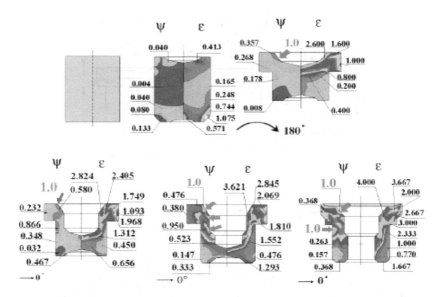

Fig. 8.2. Fields of distribution of cumulated plastic strain intensity ε (right hand half of the section) and the probability of fracture ψ (left half) in the volume of the deformed blank in the passes.

pass. In addition, the strain gradients are large. For example, at the second pass, ε varies from 0.2 to 2.6 (Fig. 8.2).

The results obtained during computer simulation allowed us to outline ways to eliminate macrocracks and rationalize the technological process from the point of view of the quality of metal products.

In order to reduce the unevenness of the deformation and its maximum values in hazardous locations, the diameter of the core part of the part was increased (starting from the first pass), the thickness of the flange and the depth of the hole under the hole were reduced (Fig. 8.3). In addition, the angle in the design was increased from 30° to 45°, which should reduce the rigidity index of the stressed state K in this region and, accordingly, increase the plasticity of the metal.

Simulation of the rationalized process of stamping the sleeve (see Fig. 8.4) showed that the unevenness of the deformation and the maximum strain values decreased. The degree of damage (the probability of macrofracture ψ) in the entire volume and at all passes is less than 1.

The maximum value of $\psi = 0.7$ is observed at the last pass in the lower upper corner of the flange.

Mathematical Formulation and Examples 221

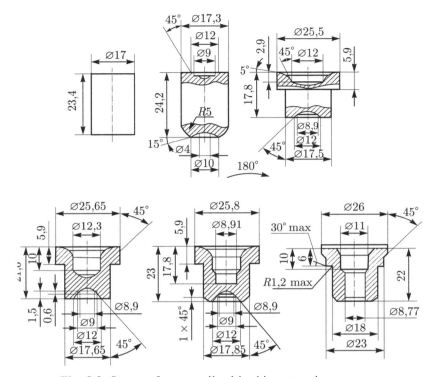

Fig. 8.3. Stages of a streamlined bushing stamping process.

After obtaining an experimental batch of suitable parts and testing them, the process was introduced into production.

Another example of improving CV~S technology on the basis of mathematical modelling using the physical and mathematical theory of plasticity is the technology of manufacturing a bolt made of steel 20G2R.

Stamping stages include (Fig. 8.5): the calibration of the rod by drawing and the length of the original workpiece of diameter 11.8 and length 82 mm; a set of metal under the head of the bolt and head upsetting in the second and third position with simultaneous reduction of the rod under the thread rolling; formation in the head of the internal hexagon on the fourth upset pass.

The technology was used in the factory, but had a significant drawback, consisting in low durability of the punches, which form a key hole in the bolt head in the fourth upset of the landing. The durability of punches at this position (the number of quality parts produced in this stamp) was 3–4 thousand pieces, and in other positions – 15–16 thousand.

222 *Physico-Mathematical Theory of High Irreversible Strains*

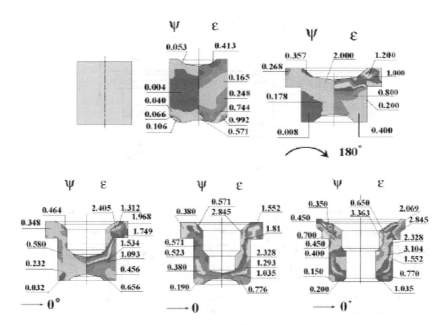

Fig. 8.4. Fields of distribution of the intensity of cumulated plastic strain and probability of fracture in the volume of the deformed blank in stages of the rationalized stamping process.

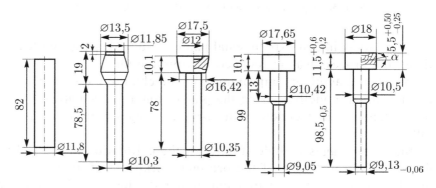

Fig. 8.5. Stages of the CVS of a bolt.

The task was to certify the technology in order to find out the reason for the low punch resistance and its possible elimination.

The mathematical formulation of the boundary value problem of the physical and mathematical theory of plasticity was formulated in a manner analogous to that described above, in addition to the absence of equations for calculating the deformation damage of a metal.

The plastic strain intensities accumulated in the passes are shown in Fig. 8.6. It can be seen that with a multi-stage CVS the strain is

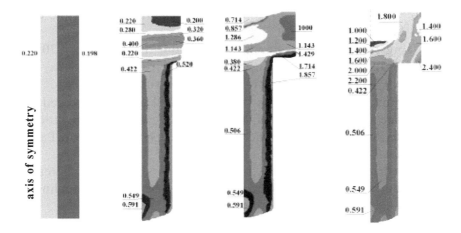

Fig. 8.6. Fields of cumulated strain in stages of CVS of a bolt (taking into account the axial symmetry halves of the axial section are shown).

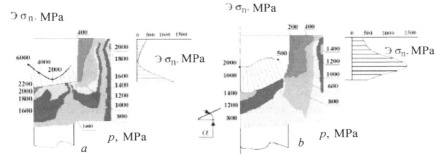

Fig. 8.7. Fields of hydrostatic presure $p = -\sigma_0$ and curves of normal contact stresses σ_n in the fourth stamping pass at $\alpha = 8°$ (*a*) and $30°$ (*b*).

large. In the penultimate pass, the maximum strain $\varepsilon = 2.4$ (Fig. 8.6). This corresponds to a relative elongation of about 200% when the specimen is uniaxially stretched. Such a large value of deformation without fracture during cold deformation of the steel is achieved due to the high positive hydrostatic pressure $p = -\sigma_0$ in the deformation zone, which reaches (2000–2200) MPa.

Figure 8.7 shows hydrostatic pressure fields p and diagrams of normal contact stresses σ_n at the end of the punch that forms the hole and on the surface of the matrix. With allowance for symmetry, the fields p and the diagrams σ_n on the right half of the section of the workpiece are given. An analysis of these results shows that the low punch resistance is due to the high values of normal contact stresses σ_n at the end of the punch when the hole is squeezed out. Local stresses in the centre of the end face of the punch reach $\sigma_n =$

6000 MPa (Fig. 8.7 a), and the average specific force at the end is 4000 MPa.

Using the numerical simulation method, the angle $\alpha = 30°$ was found (Fig. 8.7 b) (according to the standard technology $\alpha = 8°$), at which σ_n^{max} on the punch reaches the smallest value of 2000 MPa, provided that σ_n^{max} on the matrix wall does not exceed 1400 MPa (see the right diagram in Fig. 8.7 b). The specific force of deformation at the end of the punch is reduced to 1250 MPa, i.e., 3.2 times.

Industrial tests of punches with $\alpha = 30°$ showed an increase in resistance to 14–15 thousand stamped bolts. The new technology was introduced into production.

The following example of the application of physical and mathematical theory plasticity is a theoretical study of the process of plastic structure formation of aluminium AD1 deformed according to the 'hourglass' scheme (Fig. 3.20) at $T = 20°C$. The results of the experimental study of the process are described and discussed in section 3.2.2.

Taking into account the alternating character of the deformation, the investigation was carried out on the basis of the theory of a large cyclic and close to it deformation with large strains in cycles and accumulated over several cycles. The mathematical model of the process included the equations (7.6), (7.7), (5.1), (5.5). The parameters of the equations were calculated from (5.3), (5.4) and by the method described in Sections 4.3 and 5.2 on the basis of basic experiments. The dynamics of the change in the characteristics of the structure during deformation was calculated from (6.2) and (6.1).

The chemical composition and mechanical properties of the material in the state of delivery are given in the tables.

Calculations were carried out at the following values of constants and model parameters: $G = 26\ 000$ MPa; $b = 3 \cdot 10^{-8}$ cm; $m = 3.1$; $B = 10.0$; $\rho_{s0} = 8,354 \cdot 10^9$ cm^{-2}; $\lambda_c = 18.1 \cdot 10^{-4}$ cm; $A = 2.27 \cdot 10^{10}$ cm^{-2}.

Numerical implementation of the model was carried out in the DEFORM-2D software environment. The planar problem was solved. Calculation schemes for the beginning and end of the first processing cycle, compiled in accordance with the deformation scheme shown in Fig. 3.20, are shown in Fig. 8.8. Taking into account the symmetry, right half of the axial section of the workpiece with the matrix is shown hereinafter.

Table 8.1. Chemical composition of aluminium grade AD (wt.%)

Al, not less than	Impurities, no more than						
	Fe	Si	Cu	Mn	Zn	Ti	Mg	other
99.3	0.3	0.3	0.05	0.025	0.1	0.15	0.05	0.05

Table 8.2. Mechanical properties of aluminium AD1

Material	Conditional yield strength σ_{02}, MPa	Ultimate strength σ_2, MPa	Relative elongation to fracture δ, %	Relative reduction in area to fracture ψ, %
AD1	88	100	29	92

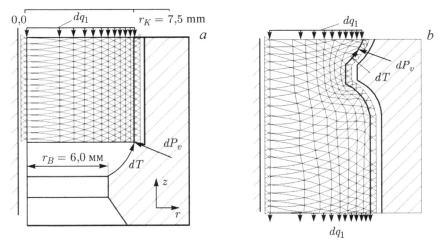

Fig. 8.8. Calculation schemes of the beginning (*a*) and end (*b*) of the first cycle of workpiece processing.

The following boundary conditions were adopted. The kinematic boundary conditions in the form of equal small displacements d_{q1} of the boundary node points were set on the contact surface of the deformable workpiece with the punch (Fig. 8.8). On the contact surface of the workpiece with the matrix, static boundary conditions were set in the form of forces dP_v directed along the normal to the surface at a given point and ensuring non-penetration of the contact nodes in the matrix body. The workpiece slides along the surface of the matrix, the contact friction being taken into account by the Coulomb relationship: $dT = \mu \cdot dP_v$, where dT is the specific frictional force, μ is the friction coefficient of 0.12.

Figure 8.9 shows the intensity distributions of the accumulated plastic strain ε (field of ε) in the axial section of the workpiece and

226 *Physico-Mathematical Theory of High Irreversible Strains*

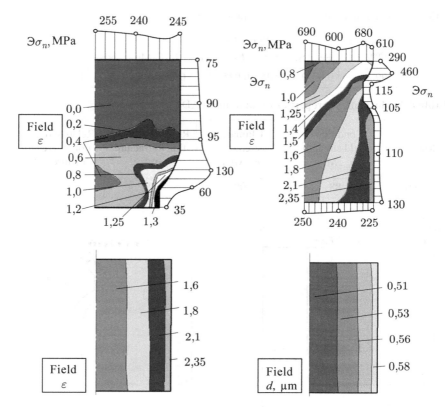

Fig. 8.9. Distribution of accumulated intensity of plastic strain ε and diagrams of normal contact stresses Эσ$_n$ at the beginning of the first processing cycle (*a*) and at the end of the second processing cycle (*b*); the distribution of the intensity of plastic deformation ε after removal of the billet after two processing cycles from the die (*c*) and the linear grain size *d* (*g*).

the diagrams of normal contact stresses $-\sigma_n$ at the end of the punch and the lateral surface of the matrix at the beginning of the first cycle (Fig. 8.9 *a*) and at the end of the second processing cycle (Fig. 8.9 *b*), as well as the distribution of ε after extrusion (removal) of the workpiece from the matrix (Fig. 8.9 *c*). The distribution of the linear grain size *d* (field *d*) in the billet treated with two cycles is shown in Fig. 8.9 *d*.

The dependence of the scalar dislocation density on the intensity of the accumulated plastic strain, calculated from equation (6.2), is shown in Fig. 8.10.

The monotonic increase in the dislocation density ρ to ε = 3.0 correlates with the monotonic hardening (Fig. 5.5, curve 2). With the dependence ρ (ε) on the plateau, the dependence σ (ε) stabilizes.

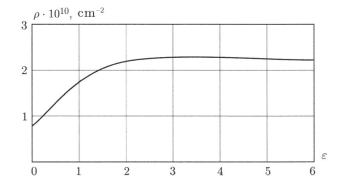

Fig. 8.10. Dependence of the dislocation density on the accumulated intensity of plastic strain in aluminium AD1.

Comparison of the results of the theoretical study with the experimental ones described in 3.2.2 shows that a theory with a satisfactory engineering accuracy predicts the size of the grains during the plastic structure formation of metals (PSM) and the value of the accumulated plastic strain ε.

Let us turn to the theoretical estimates of the grain refinement limit d^* by deformation methods. It is known that the maximum dislocation density that can be obtained in metals under cold deformation in local volumes of material ρ_{max} is 10^{12} cm^{-2} [68]. Substituting this value into formula (6.1) and assuming that all these dislocations went to the formation of deformation boundaries of the fragments, we obtain d^* equal to 0.1 μm (100 nm). This is the limiting estimate.

It is also known that plastic deformation by the mechanism of dislocation sliding is inherently uneven and the greater the volume of metal is covered by deformation, the higher its unevenness. Therefore, ρ_{max} is achieved in the localization of deformation.

In the technological processes of metal working with pressure, the unevenness of the deformation increases sharply due to the presence of contact friction forces between the workpiece being machined and the tool.

It was noted above that the softening effect is observed in the case of cyclic deformation with large strains in half-cycles, (Figs. 5.3, 5.4 and 5.5).

Theoretical calculations of the limiting dislocation density, performed for different metals using Eq. (6.2), showed that, because of the softening, it does not exceed 10^{11} cm^{-2}. Substituting this value in (6.1), we obtain d^* equal to 0.316 μm. This value of the grain refinement limit in the process of PSM corresponds to most of the experimental data published in the literature.

It follows from the above analysis that the formulation of the problem of obtaining nanostructured bulk structural material by deformation methods has no theoretical and experimental justification for today.

The problem of obtaining a microcrystalline structural material with d equal to 0.3 μm or more, which, in comparison with the existing coarse-grained materials, possesses an increased level of the characteristics of statistical, dynamic and fatigue strength and satisfactory plasticity characteristics is the real, theoretical basis and experimental justification. This metal can be used as a structural metal and can also be processed by the method of pressure metal working with pressure the under conditions of structural superplasticity [217].

Conclusion

The physical and mathematical theory of dislocation plasticity, creep, stress relaxation, ductile fracture, long-term strength and the evolution of the microstructure of metals in these processes are developed on the basis of combining micro and macro-representations about irreversible deformation and methods of its microdescription.

Unlike the classical phenomenological theories of these processes, the physico-mathematical theory consistently takes into account the loading history and evolution of the structure, and also gives a physical interpretation of the mechanical effects.

The theory unites two directions in the study of irreversible strains of metals – the physics and mechanics of strength and plasticity and extends the possibilities of design calculations in the creation of new technology.

References

1. Saint-Venant B., Liouville J. Math. 1871. V. 16. P. 308–316, 373–382.
2. Kolmogorov V.L. Mechanics of metal forming. Tutorial for universities. 2nd ed., Ekaterinburg: Ural State Technical University - UPI, 2001. - 836 p.
3. Landau L.D., Lifshits E.M. Theoretical Physics: Textbook. In 10 volumes. V. 1. Mechanics. Moscow, Nauka, 1988. - 216 p.
4. Il'yushin A.A., Continuum mechanics. Textbook. Moscow, Publishing House of Moscow University, 1978. - 287 p.
5. Il'yushin A.A., Plasticity. Moscow, Publishing House of the Academy of Sciences of the USSR, 1963. - 271 p.
6. Ivlev D.D., Bykovtsev G.I. Theory of hardening plastic body. Moscow, Nauka, 1971..
7. Malinin N.N., Applied theory of plasticity and creep. Textbook for university students. Moscow, Mashinostroenie, 1975.
8. Ishlinsky A.Yu., Ivlev D.D., Mathematical theory of plasticity. Moscow, FIZMATLIT, 2001..
9. Mathematical encyclopedic dictionary. Chief. ed. Prokhorov Yu.V., Moscow, Sov. entriklopediya, 1988.
10. Zubchaninov V. G. ,Stability and plasticity. In 2 volume. V.2. Plasticity theory, Moscow, FIZMATLIT, 2008.
11. Reiss E. Accounting for elastic deformation in the theory of plasticity. In the book: Theory og plasticity. Moscow, Izd. foreign lit., 1948, pp. 206–222.
12. Prandtl L., in: Proceedings of 1-st Jnternational congress of applied mechanics. Delft, 1924. P. 43–54.
13. Saint-Venant B. On the establishment of the equations of internal motions in solid plastic bodies beyond elasticity. In the book: The theory of plasticity. Moscow, Izd. foreign Lit., 1948. P. 11-19.
14. Saint-Venant B. Differential equations of internal motions in solid plastic bodies, and the boundary conditions for these bodies. In the book: Theory of plasticity. Moscow, Izd. Foreign Lit., 1948. p. 24–33.
15. 15. Levi M., In the book: The theory of plasticity. Moscow, Izd. Foreign Lit., 1948. p. 20–23.
16. Mises R., In the book: Theory of plasticity. Moscow, Izd. Foreign Lit., 1948. pp. 57–69.
17. Ishlinsky A.Yu., Ukrainian Mathematical Journal. 1954. Vol. 6, No. 3. Pp. 314–324.
18. Prager A., Hodge F. G. The Theory of Ideally Plastic Bodies. Moscow, Izd. Foreign Lit., 1956.
19. Kadashevich Yu.I., Novozhilov V.V., PMM. 1958. Vol. 22, No. 1. Pp. 78–89.
20. Calculations of engineering structures using the finite element method:. Reference book V.I. Myachenkov, V.P. Maltsev, V.P. Mayboroda and others; Moscow, Mashinostroenie, 1989.
21. Armstrong P. J., Frederick C.O., A Mathematical Representation of the Multiaxial Bauschinger Effect. (Central Electricity Generating Board Report No / RD / BN / N 731). Berkeley Laboratoies, R & D Department. Ca. (1966).
22. Dafalias Y.F., Popov T.T., Acta Mechanica, 1975. 21. P. 173–192.

23. 23. Chaboche J.-L., International Journal of Plasticity. 1991. 7. P. 661–678.
24. Dovnorowski W., Persyna P., Transactions of ASME. J. Eng. Materials and Technology. 1999. 121. P. 210-220.
25. Mroz Z., J. Mech. Phys. Solids. 1967. 15. P. 163-175.
26. Ohno N., Journal of Appied Mechanics, ASME.
27. Krieg R.D., Journal of Applied Mechnics. ASME. 1975. E 42. p. 641–646.
28. Chaboche J.-L., International Journal of Plasticity. 1989. 5. P. 247–302.
29. Voyiadjis G. Z., Abu Al-Rub P.K., International Journal of Plasticity. 2003. 19. P. 2121–2147.
30. Bondar V.S., Inelasticity. Variants of the theory. Moscow, FIZMATLIT, 2004..
31. Danilov V.M., Izv. AN SSSR. Mekh. Tverd. Tela .1971. No. 6. Pp . 146–150.
32. Danilov V.L., Izv. VUZ, Mashinostroenie. 1972. No. 4. Pp. 10–16.
33. Greshnov V.M., et al., Kuznechno. Shhtamp. Proizvod. 2001. No. 8. Pp. 33–37.
34. Segal V.M., et al., The processes of structure formation of metals. Minsk, Nauka i tekhnika, 1994.
35. Likhachev V.A., Malinin. V.G. Structural-analytical theory of strength. St. Petersburg, Nauka, 1993..
36. Tretyakov A.V., Zyuzin V.I., Mechanical properties of metals and alloys in pressure treatment. Directory. Moscow, Metallurgiya, 1973.
37. 37. Suyarov D.I., Lel R.V., Gilevich F.S. Hardening and softening of metals and alloys during hot plastic deformation. Gorky: GPI, 1975.
38. Polukhin P.I., et al., Resistance to plastic deformation of metals and alloys. Moscow, Metallurgiy, 1983.
39. Mazur V.L., Khizhnyak D.D., Izv. SSSR, Metally, 1991. No. 5. Pp. 148-154.
40. Pozdeev A.A., et al., Application of the theory of creep in metal forming. Moscow, Metallurgiya, 1973.
41. Rabotnov Yu.N., Creep structural elements. Moscow, Nauka, 1966.
42. Kennedy A.J. Creep and fatigue in metals. Moscow, Metallurgiya, 1965.
43. Garofalo F., Laws of creep and long-term strength of metals and alloys. Moscow, Metallurgiya, 1968.
44. Cadek I., Creep of metallic materials. Translated from Czech. Moscow, Mir, 1977.
45. Poirier J.P., High-temperature plasticity of crystalline bodies.Translated from French Moscow, Metallurgiya, 1982.
46. Pezhina P. The main issues of viscoplasticity. Moscow, Mir, 1968..
47. Il'yushin A.A., Pospelov, I.I., Inzhener. Zhurnal, 1964, V. IV, issue 4. pp. 697–704.
48. Kachanov L.M., Theory of creep. Moscow, Fizmatgiz, 1960.
49. Malinin N.N., Basics of creep calculations. Moscow, Mashgiz, 1948.
50. Sosnin O.V., Probl. prochnosti, 1973, No. 5. Pp. 45–49.
51. Shesterikov S.A., Izv. AN SSSR. Mekhanika i mashinostroenie. 1959. No. 1. P. 131.
52. Lokoshchenko A.M., Mekhanika Tverd. Tela. 2012. No. 3. Pp. 116–136.
53. Lokoshchenko A.M., Prikl. Mekh. Tekhn. Fizika. 2012. V. 53, No. 4. Pp. 149–164.
54. Rumer Yu.B., Ryvkin M.Sh., Thermodynamics, statistical physics and kinetics. Moscow, Nauka, 1997.
55. Kachanov L.M. Izv. AN SSSR. Otd tekhn. nauk. 1958. No. 8. Pp. 26–31.
56. Rabotnov Yu.N., On the mechanism of long-term destruction, in: Questions of strength of materials and structures. Moscow, Publishing House of the Academy of Sciences of the USSR, 1959. P. 5–7.
57. Levitas V.I., Large elastic-plastic deformation of materials at high pressure. Kiev, Naukova Dumka, 1987.

References

58. Kovtanyuk L.V., Mathematical model of large elastic-plastic strains and patterns of formation of residual stress fields in the vicinity of inhomogeneities of materials: Dissertation, Vladivostok, 2006.
59. Chaboce J.L., International Journal of Plasticity. 2008. No. 24. P. 1642-1693.
60. Lee E.H., Trans ASME: J.Appl. Mech. 1969. V. 36, No. 1. P. 1–6.
61. Bykovtsev G.I., Shitikov A.V. Dokl. AN SSSR. 1990. V. 311, No. 1. Pp. 59–62.
62. Myasnikov V.P., Vestn. DVO RAN. 1996. No. 4. Pp. 8–13.
63. Burenin A.A., Kovtanyuk L.V., Mazelis A.L., Prikl. Mekh. Tekh. Fiz. 2010. V. 51, No. 2. Pp. 140–147.
64. Annin B.D., Zhigalkin V.M., The behavior of materials in complex loading conditions.. Novosibirsk: Publishing House of the Siberian Branch of the Russian Academy of Sciences, 1999.
65. Khan A.S., Liang R.,International Journal of Plasticity. 2012. No. 36. P. 113–129.
66. Freund M., Shutov A. V., Ihlemann J., International Journal of Plasticity. 2008. No. 24. P. 1642-1693.
67. Shtremel' M.A., Strength of alloys. Part 1. Lattice defects. Moscow, Metallurgiya, 1982.
68. Shtremel' M.A., Strength of alloys.. Part 2. Deformation. Moscow, MISIS, 1997.
69. Popov L. E., Kobytev V.S., Kovalevskaya, T.A. Plastic deformation of alloys. Moscow, Metallurgiya, 1984.
70. Polukhin P.I., Gorelik S.S., Vorontsov V.K., Physical basics of plastic deformation. Moscow, Metallurgiya, 1982.
71. Rybin V.V., Large plastic deformation and destruction of metals. Moscow, Metallurgiya, 1986.
72. Vladimirov V.I., The physical nature of the destruction of metals. Moscow, Metallurgiya, 1984.
73. Grechnikov F.V., Zaitsev V.M., Physical and mechanical fundamentals of deformation of materials. Samara: SGAU Publishing House, 2006.
74. Snitko N.K., Zh. Tekh. Fiz., 1948. No. 6. Pp. 863–874.
75. Batdorf S.B., Budyansky B.A., Mathematical theory of plasticity on the basis of the slip concept. Mechanics. Collection of translated foreign articles. Mir, 1962. No. 1. Pp. 135–155.
76. Rusinko K.N. Features of the inelastic deformation of solids. - Lviv, Vishcha shkola, 1986.
77. Khristianovich S.A., Shemyakin E.I., Mekh. Tverd. Tela. 1969. No. 5. Pp. 138–149.
78. Knets. The main modern directions in mathematical theory of plasticity. Riga, Zinatne, 1971.
79. Mohel A.N., Salganik R.L., Khristianovich S.A., Mekh. Tverd. Tela, 1983. No. 4. Pp. 119–141.
80. Kukudzhanov V.N., Izv. RAN, Mekh. Tverd. Tela. 2006. No. 6. Pp. 103–135.
81. Berisha B., Hora P., Wahlen A. International Journal Material Forming 2008. No. 1. P. 135–141. DOI 10. 1007 / s 12289-008-0367-7.
82. Luce R., Wolske M., Kopp R., Roters F., Gottstein G., Computational Materials Science. 2001. No. 21. P. 1–8.
83. Abed Farid H. Physically based multiscale-viscoplastic model for metals and alloys: theory and computation. http: etd.lsu.edu/docs/available/etd-07062005-094112/.
84. 84. Lin J., Dean T.A., International Journal of Mechanical Science. 2007. V. 49, No. 7. P. 609–918.
85. Storozhev M.V., Popov E.A., Theory of metal forming. Moscow, Mashinostroenie, 1971..

86. Gun G.Ya. The theoretical basis of metal forming. (Theory pf plasticity). Moscow, Metallurgiya, 1980.
87. Smirnov-Alyaev G.A. The material resistance to plastic deformation. Leningrad, Mashinostroenie, 1978.
88. Unksov E.P., et al., Theory of plastic deformation of metals. Moscow, Mashinostronie, 1983.
89. Forging and stamping: a handbook. In 4 volumes, Ed. by E.I. Semenov and others. Moscow, Mashinostroenie, 1987. V. 3. Cold volumetric stamping, ed. E.I. Semenov, 1986.
90. Forging and stamping: a handbook. In 4 volumes, Ed. by E.I. Semenov and others. Moscow, Mashinostroenie, 1986. V. 2. Hot stamping, ed. G.A. Navrotsky, 1987.
91. Smirnov O.M.. Metal processing by pressure in the state of superplasticity, Moscow, Mashinostroenie, 1979..
92. Figlin S. Z., et al., Isothermal deformation of metals, Mashinostroenie, 1978.
93. 93. Yakovlev S.P., Chudin V.N., Yakovlev S.S., Sobolev Ya.A. Isothermal deformation of high-strength anisotropic metals. Moscow, Tula, Mashinostroenie-1, Tula: Publishing house of TSU, 2003.
94. Khan A.S., Farrokh B., Takacs L., Journal Materials Science. 2008. No. 43. P. 3305–3313. DOI 10/1007 / s 10853-008-2508-2.
95. Sosnin O.V., Gorev B.V., Nikitenko A.F. Energy option of the creep theory. Novosibirsk: Inst. of Hydrodynamics, 1986.
96. Sosnin O.V., Nikitenko A.F., Gorev B.V., Prikl. Mekh. Tekhn. Fizika, 2010. V. 51, No. 4. Pp. 188–197.
97. Sosnin O.V., Lyubashevskaya I.V., Novoselya, I.V., ibid, 2008. T. 49, No. 2. Pp. 123–130.
98. Artsruni A.A., Nikitenko A.F., in: Actual Problems of Protection and Safety, 8th All-Russian conf. Technical means of countering terrorism. St Petersburg. V. 1. S. 173–180.
99. Radchenko V.P., Saushkin M.N., Gorbunov S.V., Prikl. Mekh. Tekhn. Fizika, 2014. V. 55, No. 1. Pp. 207–217.
100. Tsvelodub I.Yu., ibid, 2012. V. 53, No. 4. Pp. 98–101.
101. Simonyan A.M., Mekh. Tverdogo Tela, 2011. No. 6. Pp. 131–138.
102. Radchenko V.P., Saushkin M.N., Goludin E.P., Prikl. Mekh. Tekhn. Fizika,2012. V. No. 2. Pp. 167–174.
103. Kukudzhanov V.N. Computer simulation of deformation, damage and destruction of inelastic materials and structures: Textbook, Moscow, MIPT, 2008..
104. Bogatov A.A., Mechanical properties and models of the destruction of metals. Ekaterinburg: GOU VPO USTU-UPI, 2002.
105. Stepanov V.A. The role of deformation in the process of destruction of solids, in: Problems of strength and ductility of metals: (LFTI). Leningrad, Nauka, 1979. P. 10–26.
106. Skudnov V.A. Extreme plastic deformation of metals. Moscow, Metallurgiya, 1989..
107. Goldenblat I.I., Konov V.A. Strength and plasticity criteria of construction materials. Moscow, Mashinostroenie, 1968.
108. Gridnev V.N., Gavrilov V.G., Meshkov Yu. Ya., Strength and plasticity cold-formed steel. Kiev, Naukova Dumka, 1974.
109. Zhurkov S.N. Fizika Tverdogo Tela, 1983. V. 25, No. 10. pp. 3119–3122.
110. Kiyko I.A., The theory of destruction in the processes of plastic flow, in: Metal forming. - Sverdlovsk: UPI, 1982. - p. 27.

111. 111. Lebedev, A.A., Chausov, N.G., Boginich, I.O., Probl. Prochn. 1997. No. 3. Pp. 55–69.
112. Mikhalevich V.M., Metally. 1991. No. 5. Pp. 89–95.
113. Worswick M. J., Pick R. J., J. Mech. and Phys. 1990. V. 38, No. 5. P. 601-625.
114. Chen O, Kobayashi M., ASME, Design. 1975. V. 101, No. 1. Pp. 114–123.
115. Fedorov V.V., Romashov R.V., Khachaturian S.V., Korshunov V.Ya. Non-equilibrium phase transitions in the destruction of metals and alloys. Lyubertsy: VINITI, 1988.
116. Getsov L.B., Probl. Mashinostr. Nadzhn. Mashin. 2001. No. 5.Pp. 49–55.
117. Smirnov S.V. Deformability and damage to metals during the pressure treatment. Dissertation, Ekaterinburg: Ural Branch of the Russian Academy of Sciences, 1998..
118. Rabotnov Yu.N. Mechanics of a deformable solid. Moscow, Nauka, 1979.
119. Lokoshchenko A.M., Prikl. Mekhanika Tekhn. Fiz. 2014. V. 55, No. 1. Pp. 144–165.
120. Troshchenko V.T., Deformation and destruction of metals during multi-cycle loading. Kiev: Naukova Dumka, 1981.
121. Zavoychinskaya E.B., Mekh. Tverd. Tela. 2012. No. 3. Pp. 54–77.
122. Lokoshchenko A.M., Mekh. Tverdogo Tela. 2010. No. 4. Pp. 164–181.
123. Jinn J. T. DEFORM, Users Meeting Presentation, Scientific Forming Technologies Corporation, Columbus, OH, May 2002.
124. Czoboly E., Havas I., Gillemot F. Proc. of Symp. on Absorbed Spec. Energystrain energy criterian. - Budapest: Academ. Kiado. 1982. - P. 107-129.
125. Li Q.M., Int. J. Solids and Struct. 2001. V. 38, No. 38–39. P. 6997-7013.
126. Latham D. J., Cocroft M.G., The Effect of Stress System on the Workability of Metals. National Engineering Laboratory, Report No. 216, Feb. 1966.
127. Li H., Fw M.W., Lu J., Yang H., International Journal of Plasticity. 2011. No. 27. P. 147–180.
128. Bai Y., Wierzbicki T., International Journal of Plasticity. 2008. No. 24. P. 1071-1096.
129. Zadpoor A.A., Sinke J., Benedictus R., International Journal of Plasticity. 2009. No. 25. P. 2269–2297.
130. Sun X., Choi K. S., Liu W.N., Khaleel M.A., International Journal of Plasticity. 2009. No. 25. P. 1888–1909.
131. Brnig M., Gerke S., International Journal of Plasticity. 2011. No. 27. P. 1598–1617.
132. Khan A.S., Liu H., International Journal of Plasticity. 2012. No. 35. P. 1–12.
133. Mae H., Teng X., Bai Y., Wierzbicki T., Materials Science and Engineering: A.2007. 459. 156–161.
134. Gurson A.L. Trans ASME. J. Eng-ng. Materials and Technology. 1977. V. 99, No. 1. P. 2–15.
135. Chu C.C., Needleman A., Trans ASME. J. Eng-ng. Materials and Technology. 1980. V. 102, No, 3. P. 249–256.
136. Tvergaard V., Intern. J. Fracture Mechanics. 1981. V. 17, No. 4. P. 389–407.
137. Kittel H., Introduction to solid state physics. Moscow, Nauka, 1978.
138. Nabarro F.R. (Ed.) Dislocations in Solids. V. 1-6. - Amsterdam: North-Holland Publ., 1983.
139. Friedel S. Dislocations. Translated from English. Moscow, Mir, 1967.
140. Hirt J., Lote I., Theory of dislocations. Translated from English. Moscow, Atomizdat, 1972.
141. Kosevich A.M. Physical mechanics of real crystals. Kiev, Naukova Dumka, 1981.
142. Novikov I.I., Portnoy V.K., Superplasticity with ultrafine grains. Moscow, Metallurgiya, 1981.
143. Greshnov V.M., Izv. AN SSSR, Metally. 1989. No. 2. Pp. 53–62.

References

144. Greshnov V.M., ibid, 1991. No. 1. Pp. 156–160.
145. Greshnov V.M., Izv. AN SSSR, Metally. 1991. No. 2. Pp. 141–145.
146. Becker R., Physikalische Zeitschrift. 1925. V. 26. P. 919–925.
147. Becker R., Zeitschrift fuer Technische Physik. 1926. V. 7. P. 547.
148. Panin V.E., et al., Izv. VUZ. Fizika. 1982. No. 6. P. 5–27.
149. Rybin V.V., Zolotorevsky N.Yu., Zhukovsky I.M. Fiz. Met. Metalloved. 1990. No. 1. Pp. 5–26.
150. Rybin V.V., Voprosy materialoved. 2002. No. 1. Pp. 11–33.
151. Panin V.E., Grinyaev Yu.V., Elsukova T.F., Izv. VUZ. Fizika. 1981. No. 11. P. 82–86.
152. Rybin V.V., Perevezentsev V.N., Pism'ma v Zh. Teor. Fiz. 1981. V. 7, No. 19. Pp. 1203–1205.
153. Perevezentsev V.N., Rybin V.V., Orlov A.N.. Poverkhnost'. 1982. No. 6. Pp. 134–142.
154. Segal V.M., Metally. 2004. No. 1. Pp. 5–13.
155. Greshnov V.M., Puchkova I.V., Prikl. Mekh. Tekhn. Fiz., 2010. T. 51, No. 2. Pp. 160–169.
156. Astanin V.V., Kaibyshev O.A., Faizova S.N., Scripta Met. et Mater. 1991. V. 25, No. 12. P. 2663–2668.
157. Greshnov V.M., in:Abstracts of the reports of the All-Union Symposium 'Questions of the theory of plasticity in modern technology'. Moscow, Moscow State University, 1985.
158. Perevezentsev V.N., Rubin V.V., Chuvil'deev V.N. Acta Met. Mater. 1992. V. 40, No. 5. P. 887–894.
159. Higashi K., Mater. Sci. Forum. 1994. 170-172. P. 131–134.
160. Kaibyshev O. A., Valiev R. Z., Emaletdinov A.K., DAN SSSR. 1984. V. 279. P. 369-372.
161. Greshnov V.M., Ivanov M.A., Metallofizika. 1993. V. 15, No. 7. S. 312.
162. Polyakov A.A.. Metalloved. Term. Obrab. Met. 1994. No. 3. P. 18–21.
163. Greshnov V.M., Pis'ma v Zh. Teor. Fiz. 1991. V. 7, No. 14. P. 5–9.
164. Emaletdinov A.K., Fiz. Tverd. Tela. 1999. V. 41, no. 10. pp. 1772-1777.
165. Emaletdinov AK, Pis'ma v Zh. Teor. Fiz. 1998. V. 24, No. 13. Pp. 43–47.
166. Panin V.E., et al., Izv. VUZ. Fizika. 1982. No. 12. Pp. 5–28.
167. Egorushkin V.E., et al., ibid, 1987. No. 1. P. 9–32.
168. Fedorov V.V., et al., The structural-energy state in extremely deformed. metals and alloys. Lyubertsy, VINITI, 1989..
169. Greshnov V.M., Golubev OV, Rtishchev A.V., Kuznechno-shtampovochnoe proizvodstvo. 1997. No. 2. Pp. 8–10.
170. Mishin O.V., Alexandrov J.V., Greshnov V.M., Golubev O.V., Valiev R. Z., in: Proceeding of the International conference 9th Symposium on Metallography. Tatransha Lomnica – Stara Lesna, Slovakia, 26–28 April, 1995. P. 315–318.
171. Grabovetskaya G.P., et al., Fiz. Met. Metalloved. 1997. V. 83, No. 3. pp. 112–116.
172. Smirnova N.A., et al., ibid, 1986. V. 61, No. 6. P. 1170–1177.
173. Pavlov V.A., et al., ibid, 1984. Vol. 58, No. 1. pp. 177–184.
174. Lyakishev N.P., Metally. 2004. No. 1. P. 3-4.
175. Valiev R.Z., ibid, 2004. No. 1. P. 15–21.
176. Valiev R.Z., Aleksandrov I.V., Nanostructured materials obtained severe plastic deformation. Moscow, Logos, 2000.
177. Kopylov V.I., Chuvildeev V.N., Metally. 2004. No. 1. Pp. 22–35.
178. Moreno-Valle, et al., Materials Letters. 2011. V. 65. P. 2917–2919.
179. Marco J. S., Acta Materialia. 2009. V. 57. P. 5796–5811.

References

180. Stepanov N.D., et al., Materials Science and Engineering A. 2012. P. 105–115.
181. Perevezentsev V.N. Theory of high-rate superplasticity. Fiz. Met. Metalloved. 1997
182. Nigmatullin R.I., Kholin N.N., Mekh. Tverd. Tela. 1974. No. 4. P. 131.
183. Greenberg B.A., et al., Phys. Stat. Sol. (a) 1976. V. 38. P. 653–662.
184. Grinberg B.A., Ivanov M.A., Fiz. Met. Metalloved. 1994. V. 78, No. 3. Pp. 3–32.
185. Ivanov M.A., et al., ibid, 1998. V. 86, No. 3. P. 24–38.
186. Vladimirov V.I., Fizika Tverd. Tela. 1970. V. 12, No. 6. P. 1593–1596.
187. Betekhtin V.I., et al., Probl. Prochnosti. 1979. No. 7. P. 38–45.
188. Betekhtin V.I., et al., ibid, 1979. No. 8. P. 51–57.
189. RybinV.V., Vergazov A.N., Fiz. Met. Metalloved. 1977. V. 49, No. 4. P. 858–865.
190. Rybin V.V., Zisman A.A., Zhukovsky I.M., Probl. Prochnosti. 1982. No. 12. P. 10–15.
191. Greshnov V.M. Fundamentals of physical and mathematical theory of irreversible deformation of metals. Structural and phenomenological approach. Saarbruecken, Deutschland: Palmarium Academic Publishing, 2013. 102 p.
192. Greshnov V.M., Mechanics of Solids. 2011. V. 46, No. 4. P. 544–553.
193. Greshnov V.M., Safin F.F., Greshnov M.V. Probl. Prochnosti.. 2002. No. 6. P. 107–115.
194. Greshnov V.M., Safin F.F., Greshnov M.V., ibid, 2003. No. 1. P. 87–97.
195. Greshnov V.M., Lavrinenko Yu.A., Napalkov A.V., Kuznechno-shtampovochnoe proizvodstvo. 1998. No. 5. P. 3–6.
196. Greshnov V.M., Lavrinenko Yu.N., Napalkov A.V., Probl. Prochnosti. 1999. No. 1. P. 68–76.
197. Greshnov V.M., Lavrinenko Yu.N., Napalkov A.V., ibid. 1999. No. 2. Pp. 74–84.
198. Greshnov V.M., Botkin A.V., Napalkov A.V., Lavrinenko Yu.A., Kuznechno-shtampovochnoe proizvodstvo. 2011. No. 10. Pp. 34–39.
199. Korbel A., Richert M., Acta Metallurgica. 1985. V. 33. P. 1971-1978.
200. Czichos H., Saito T., Smith L. Springer Handlook of Materials Measuremen Methods. Springer, 2006.
201. Araujo M.C., Non-Linear Kinematic Hardening Model for Multiaxial Cyclic Plasticity. Thesis 2002.
202. Greshnov V.M., Pychkova I.V., Journal of Applied Mechanics and Technical Physics. 2008. V. 49, No. 6. P. 1021-1029.
203. Lokoshchenko FM, Nazarov V.V., Platonov D.O., Shesterikov S.A. Mekh. Tverd. Tela. 2003. No. 2. P. 139–149.
204. Dacheva MD, Shesterikov S.A., Yumasheva M.A., ibid, 1998. No. 1. P. 44–47.
205. Yao Hua-Tang, Xuan Fu-Zhen, Wang Zhengdong, Tu Shan-Tung. Nuclear Engng Design. 2007. V. 237. P. 1969-1986.
206. Leckie F.A., Hayhurst D. R., Proc. Roy. Soc. London Ser. A. 1974. V. 340, No. 1622. P. 323–347.
207. Othman A.M., Dyson B.F., Hayhurst D. R., Lin J. Acta Metallurgica et Materialia. 1994. V. 42, No. 3. P. 597–611.
208. Greshnov V.M., Shaikhutdinov R.I., Physical and phenomenological model dislocation creep of metals. Vestnik UGATU. 2013. V. 17, No. 1 (54). P. 33–38.
209. Sorokin V. G., Handbook of steels and alloys, Moscow, Mashinostroenie, 1989.
210. Greshnov V.M., Botkin A.V., Napalkov A.V., Lavrinenko Yu.A., Kuznechno-shtampovochnoe proizvodstvo. 2001. No. 10. Pp. 34–39.
211. Degtyarev V.P., Plasticity and creep of engineering constructions. Moscow, Mashinostroenie, 1967.

References

212. Malygin G.A., Fiz. Tverd. Tela, 2006. V. 48, No. 4. Pp. 651–657.
213. Greshnov V.M., Oshnurov A.V., Alentev V.V., Artyukhin V.I., Kuznechno-shtampovochnoe proizvodstvo. 2004. No. 2. Pp. 33–37.
214. Segerlind L. Application of the finite element method. Moscow, Mir, 1979.
215. Zienkiewich O., Method of finite elements in engineering. Moscow, Mir, 1975.
216. Greshnov V.M., Journal of Applied Mechanics and Technical Phisics. 2008. V. 49,
217. No. 6. P. 1021-1029.
218. Kaibyshev Sh. F., Utiashev, F. Z. Superplasticity, refining of structures and processing of hard-to-deform alloys. Moscow, Nauka.
219. Lepin G.F., Metal creep and heat resistance criteria. Moscow, Metallurgiya. 1976.
220. Bulygin I.P., Atlas of tension diagrams at high temperatures, creep curves and long-term strength of steels and alloys for engines. Moscow, Oborongiz, 1957.

Index

B

Bogatov A.A. 61, 63

C

CAE programs 16
condition
  Drucker stability condition 157
  Garson plasticity condition 76
  Huber–von Mises plasticity condition 13, 18
  Lee condition 42
  von Mises plasticity condition 154
crack
  Griffith crack 60
creep
  Coble diffusion creep 30, 105
  Harper–Dorn creep 124
  Nabarro–Herring creep 30. 123, 124
criteria
  Rice & Tracy, Osakada deformation criteria 68
criterion
  Cockroft–Latham criterion 67
  force criterion of fracture 59
  Freudenthal energy criterion 67
  Oyane, Ayada and Brozzo criteria 68
  Zhao & Kuhn force criterion 68
curve
  creep curve 27, 28, 30, 33, 37, 40
  isochronous creep curves 175
  relaxation curve 28, 39

D

damage
  deformation damage 52, 57, 59, 61, 62, 66, 69, 78
Davenport W.H. 34

defects
  equilibrium lattice defects 80
DEFORM program 153
density
  scalar dislocation density 83, 169, 206, 207, 226
deviator
  deviator of additional stress 18
  deviator of the active stress 18
diagram
  plasticity diagram 58, 59
dislocations 79, 80, 82, 83, 84, 87, 88, 89, 90, 91, 92, 93, 94, 95, 96, 97, 99, 100, 101, 102, 103, 104, 105, 108, 117, 118, 119, 120, 121, 122, 124, 125, 126, 127, 131, 132, 133, 134, 135, 136, 138, 139, 140, 141, 142, 143, 144, 146, 148, 152, 164, 165, 170, 171, 172, 173, 186, 188, 189, 194, 199, 202, 206, 227
  forest dislocations 93, 139, 186
  mobile dislocations 88, 89, 90, 92, 93, 118, 119, 120, 139, 140, 144, 164
  stationary dislocations 87, 90, 93, 96, 119, 139, 140, 141, 146, 152, 164, 171, 172, 188

E

effect
  Bauschinger effect 17, 18, 19, 75, 77, 165, 166
equation
  Arrhenius equation 98
  creep state equation 52
  equation of heat conductivity 3
  Hall–Petch relation 93
  Kadashevich–Novozhilov equation 21
  Nabarro–Herring equation 124
  Orowan equation 92, 97, 118, 144
  Prandtl–Reuss equations 15
  Saint-Venant geometric relations 4
  Saint-Venant–Levi–von Mises equations 15

F

fluidity
  Huber–von Mises fluidity 5
Frank–Read sources 118, 121
function
  dissipative function 11
  loading function 5, 9, 10, 11, 12, 13, 19, 22, 25, 55

H

hardening
  anisotropic hardening 17, 18, 19, 20, 21
  dispersion hardening 126
  kinematic hardeninh 19, 21, 26, 33, 41, 45
  strain hardening 92, 93
  translational hardening 19, 24
hypersurface
  creep hypersurface 31

I

Il'yushin A.A. 12, 35
intensity
  stress intensity 5, 13, 14, 15, 49, 53, 54
Ishlinsky A.Yu. 19, 20, 21

K

Kachanov L.M. 34, 35, 36
Kadashevich Yu.O. 19, 20, 21, 23
Kolmogorov V.I. 2, 61, 63

L

law
  associated flow law 9, 10, 11, 12, 13, 19, 23, 53
  Hooke's law 5, 6, 7, 15, 38
  Schmid law 85, 86, 106
Lokoshchenko A.M. 35
Ludwik P. 34

M

macroscopic continuity 3
method
  Lagrange multiplier method 9
model
  Armstrong–Frederick model 22
  Batdorff–Budyansky model 45
  coupled plasticity models 52
  creep model 198
  discrete (atomic) model of metallic materials 79
  Garzon–Tvergaard–Nidelman model 69
  GTN model 69, 76, 77, 78
  Kadashevich–Novozhilov model 20, 23
  kinetic physical-phenomenological model of dislocation creep 170

Maxwell's body model 38
model of a large cyclic and near-plastic deformation 208
model of ductile fracture of metals under developed plastic strains 63
model of long-term strength 176, 180, 182, 203, 204
multi-surface models 21, 22, 49
stress relaxation model 183, 203
uniaxial physico-phenomenological creep model 175
Voigt body model 38
Weertman model 121

N

Nadai A. 34
Novozhilov V.V. 19, 20, 21, 23

O

Orowan E. 45

P

parameter
　Nadai–Lode parameter 58, 68
　Udquist hardening parameter 13, 36, 49, 54
Peierl's relief 84
Polanyi M. 45
polygonization 94, 96, 99
postulate
　Drucker postulate 7, 50, 52, 157
Prager W. 18, 19, 20, 21
principle
　principle of the maximum dissipation rate 10, 11
　principle of the maximum work of plastic deformation 8
　Volterra principle 37
　von Mises maximum principle 10, 44
　Ziegler principle 11

R

Rabotnov Yu.N. 26, 32, 34, 35, 36, 52, 54
relaxation
　stress relaxation 27, 28, 39, 40

S

scheme
　Polanyi–Taylor scheme 99

Soderberg C.R. 32
softening 93, 139, 140, 142, 143, 146, 147, 155, 164, 167, 227
stage
  easy slip stage 91
  multiple slip stage 91
superdislocation 95
superplastic deformation 96, 100, 103, 104, 105, 107, 108, 118
surface
  surface of the beginning of plasticity 5

T

Taylor rotation of the grain 101
tensor
  active stress tensor 74
  Almansi strain tensor 42
  annihilation tensor 73
  Cauchy stress tensor 43, 70
  plastic strain tensor 74
theory
  flow theory with isotropic hardening 15
  technical creep theories 32, 35, 54
  theory of ductile fracture of materials under developed plastic strains 59
  theory of heredity 36
thermal recovery 87

V

vacancy migration 81
volume
  activation volume 81, 87, 94, 132, 142, 144